AF649391

SOIL TESTING & ANALYSIS

Plant, Water and Pesticide Residues

NIPA® GENX ELECTRONIC RESOURCES & SOLUTIONS P. LTD.
New Delhi-110 034

www.nipaers.com
Browse, Search, Read & Buy...
Print Books
eChapters
eBooks
Publishers
Forthcoming
Subject Catalogues
New Titles
Online Resources on:
✓ Current Affairs
✓ Reasoning, Logic & Aptitude
✓ Competitive Examinations
✓ English Language Lab
✓ Writing & Pronunciation Tools
✓ Personality Development
✓ Online Programmes for Professionals
✓ Interview Preparation
Pay Using
PhonePe
Google Pay
paytm
PayPal
UPI
CC Avenue

SOIL TESTING AND ANALYSIS

Plant, Water and Pesticide Residues

PATIRAM

Principal Scientist (Soil Science)

BRAJENDRA

Scientist (Soil Science)

N. S. AZAD THAKUR

Principal Scientist (Entomology)

T. RAMESH

Scientist (Soil Science)

ICAR Research Complex for NEH Region
Umroi Road, Umiam - 793 103, Meghalaya

NIPA® GENX ELECTRONIC RESOURCES & SOLUTIONS P. LTD.

New Delhi-110 034

NIPA® GENX ELECTRONIC RESOURCES & SOLUTIONS P. LTD.

101,103, Vikas Surya Plaza, CU Block
L.S.C. Market, Pitam Pura, New Delhi-110 034
Ph : +91 11 27341616, 27341717, 27341718
E-mail: newindiapublishingagency@gmail.com
web: www.nipabooks.com

For customer assistance, please contact
Phone: + 91-11-27 34 17 17
Fax: + 91-11- 27 34 16 16
E-Mail: feedbacks@nipabooks.com

© 2023, Publisher

ISBN: 978-93-95763-98-1

All rights reserved. No part of this publication may be reproduced, stored in a retrieval system or transmitted in any form or by any means, including electronic, mechanical, photocopying recording or otherwise without the prior written permission of the publisher or the copyright holder.

This book contains information obtained from authentic and highly reliable sources. Reasonable efforts have been made to publish reliable data and information, but the author/s, editor/s and publisher cannot assume responsibility for the validity, accuracy or completeness of all materials or information published herein or the consequences of their use. The work is published with the understanding that the publisher and author/s are not attempting to render any professional services. The author/s, editor/s and publisher have attempted to trace and acknowledge the copyright holders of all material reproduced in this publication and apologize to copyright holders if permission and/or acknowledgements to publish in this form have not been taken. If any copyrighted material has not been acknowledged, please write to us and let us know so that we may rectify the error, in subsequent reprints.

Trademark Notice: NIPA®, the NIPA® logos and their presentations (the way they are written/ presented) in this book are the trademarks of the publisher and hence may not be used without written permission, if copied or used without authorization, the infringer will be prosecuted as per law.

NIPA® also publishes books in a variety of electronic formats. Some content that appears in print may not be available in electronic books, and vice versa.

Composed and Designed by NIPA®.

भाक्र अनूप
ICAR

भारतीय कृषि अनुसंधान परिषद
उत्तर पूर्वी पर्वतीय कृषि अनुसन्धान संस्थान
उमरोई रोड, उमियम (बड़ा पानी) मेघालय

ICAR Research Complex for N.E.H. Region
Umroi Road, Umiam -793103, Meghalaya

Foreword

A wake up alarm has already been sounded globally that agriculture is approaching towards its carrying capacity due primarily to the degradation in the natural resource base particularly" the soil, water and forest. In last 50 years we have experienced that almost 50% of world's forest area has reduced. Soil, which is vital for food production has been observed to be increasingly being deficient in its micronutrient contents not to mention about the decreasing trend in the organic carbon. Similarly water availability has been projected to go down in the crop field from the present level of 85% to around 70% by 2050. Continuous application of agrochemicals particularly the pesticides has been responsible to create a situation where the pesticide residues in foods are being reflected.

In view of this situation, agriculture in India is today in its cross road. Agricultural scientists are today facing the challenge of meeting the dual responsibilities i.e. to ensure food availability for 1 billion plus people from shrinking resources and the other being the challenge of providing safe food where the toxic elements should not cross the permissible limit. Such challenge can only be met if the relationship among soil and water is clearly understood and mass propagated for creating awareness among the user groups of the resources.

It is in this context that the scientists belonging to Soil Science and Entomology Divisions of the Institute have brought out a book entitled **"SOIL TESTING & ANALYSIS (Plant, Water and Pesticide Residues)"**. I am sure the book would provide an insight to their relationships.

I am confident that the readers of the book shall be immensely benefited in planning their resources management activities in order to provide sustainability to agricultural sector.

I compliment the authors for bringing out the book at an appropriate time.

(Dr. K. M. Bujarbaruah)
Director

भाकृ अनुप
ICAR

भारतीय कृषि अनुसंधान परिषद
उत्तर पूर्वी पर्वतीय कृषि अनुसन्धान संस्थान
उमरोई रोड, उमियम (बड़ा पानी) मेघालय

ICAR Research Complex for N.E.H. Region
Umroi Road, Umiam -793103, Meghalaya

Message

Soil and water are the most important natural resources that support and provide nutrients to plants and animals for our survival. It also buffers much of our waste materials helping to correct and often hide many of our mistakes and oversights. If managed properly they provide protection for our environment. The **"SOIL TESTING & ANALYSIS (Plant, Water and Pesticide Residues)"** provides timely and needed available procedures for their analysis. The book will be of immense help to subject matter specialists of KVKs, extension workers, researchers, students and personnel engaged in quantification of soil, plant, water and pesticides residues analysis. It will particularly meet the requirements of various soil testing laboratories engaged in soil and water analysis of their respective states for fertilizer recommendations and water pollution.

Authors described the chapters in such a way one can get the introduction about the subject matters for their easy understanding in actual performance of the work. I wish this book would meet the needs of those who are engaged in their respective fields.

RNPrasad

(Dr. R.N.Prasad)
Ex-Director, ICAR Research Complex for NEH Region
Umiam, Meghalaya & Ex -ADG (Soils) ICAR, New Delhi

Preface

The analysis of soil, water and plant are the essential tool for the optimization of crops productivity as well as quality of produce. Soil testing for the available nutrient status is prerequisite before the sowing of crops to predict the response of crops to added fertilizer nutrients and to avoid the excess use of nutrients. The soil analysis for the available nutrient reserve of soils alone is not sufficient to guide the nutrient status of plants. So the complementary soil and plant analysis can give almost true picture of soil nutrient status and factors affecting the nutrient availability to plants. At present more than 400 soil testing laboratories are engaged for the analysis of soils all over the country and to advice the farmers to use the fertilizers for optimum crop productivity based on their field's soil tests. In nature it is difficult to get pure water for agricultural as well as for drinking purposes. The conventional agriculture production system added additional problem to pollute water for safe use. Thus water may be saline, alkaline, acidic, turbid, pesticide residues, etc., so analysis for the both purposes is very much essential for animals including human being health.

This book will be of great help to graduate/post graduate students of agriculture, forestry and environmental sciences, subject matter specialists (soil and agronomy) of the Krishi Vigyan Kendras engaged in the transfer of technology to the farmers to optimize the productivity. This book will be of great help to those who are engaged to analyze the soil, water and plant in their laboratories for carrying out routine analysis. Some basic conversion factors have also been given for beginners and further common soil science glossary for ready reference. Authors sincerely hope that the book can play a significant role to popularize the fertilizer use on the basis of soil and plant tests for boosting the food production.

June, 2007 **Authors**

Acknowledgements

The authors sincerely express their gratitude to Dr. K.M. Bujarbaruah, Director of the institute, for his support and encouragement to write a book on soil, plant, water and pesticide residues analysis. Authors are also grateful to Dr. V. Subramaniam, Zonal Coordinator, Zonal Coordinating Unit, Zone III for encouragement and providing necessary helps. Technical support of L.Somendro Singh (Research Associate) and L.J.Bordoloi (Technical officer) are also acknowledged.

Dated: 04-03-2007 **Authors**

CONTENTS

1

Introduction

Crop production broadly depends on the fertility of the soil where a crop is raised. There are several factors like the kind and quality of seed, climate, water supply, and plant protection measures etc. that affect the crop production. Among these factors, the fertility status of the soil largely determines the ultimate yield even if all these factors are optimum for crop production. A clear distinction should be made between soil fertility and soil productivity. Soil fertility deals with the inherent capacity of soil to provide nutrients in adequate amounts and in proper balance, for the growth of specified plants when other growth factors such as light, water and temperature, and the physical condition of the soil are favourable. Soil productivity is concerned with the capability of soil to produce a specified plant or plant parts or a sequence of plants under well defined and specified systems of management inputs and environmental conditions, *viz.* climatic conditions. It does not mean that a fertile soil is always productive. A fertile soil, under a bad management system, may be unproductive for a crop. In this book, plant nutrient elements and their available forms, functions and deficiency symptoms are briefly described for the better understanding of soil testing and fertilizer recommendation.

1.1 Plant Nutrient Elements

Plants accumulate nutrient elements differently, absorb some elements in large quantity, and some others in relatively smaller quantities and completely reject some others. Based on the 'criteria of essentiality' proposed by Arnon and Stout (1939), there are sixteen elements considered as essential for the growth of higher plants. These are carbon (C), hydrogen (H), oxygen (O), nitrogen (N), phosphorous (P), potassium (K), calcium (Ca), magnesium (Mg), sulphur (S), iron (Fe), manganese (Mn), zinc (Zn), copper (Cu), molybdenum (Mo), boron (B) and chlorine (Cl). These essential nutrient elements are classified into macronutrient elements and micronutrients based on their relative abundance in plants or the quantity required by the plant.

1.1.1 Macronutrients

Macronutrients or major nutrients are required in large quantities, more than that of micronutrients. These include C, H, O, N, P, K, Ca, Mg and S. Carbon, hydrogen and oxygen constitute 90 to 95 per cent of plant dry matter weight and are taken from air and soil water. These are also called as basic nutrients. The remaining six macronutrients are further subdivided into primary and secondary nutrients. *Primary nutrients* are required in large quantities. Their widespread deficiencies occur and often corrected through application of commercial fertilizers. It includes nitrogen, phosphorous and potassium. *Secondary nutrients* are required in moderate quantities by plants, and their localized deficiencies are corrected by incidental accretions through carriers of primary nutrients like single super phosphate, ammonium sulphate, etc. It includes calcium, magnesium and sulphur.

1.1.2 Micronutrients

These are required in very small quantities, generally reported in ppm as compared with kg/ha in the case of primary and secondary nutrients but are essential as macronutrients. These include iron, manganese, zinc, copper, molybdenum, boron and chlorine. Micronutrients are further subdivided into micronutrient cations (Fe, Mn, Zn and Cu) and micronutrient anions (B, Mo and Cl) depending upon the form in which plants absorb them.

In soil, these nutrient elements occur in different forms like organic and inorganic forms. Organic forms of nutrients are, first, converted into inorganic forms by decomposition processes for the ready uptake of plants. Inorganic forms include nutrients present in the soil solution, exchangeable sites of clays and organic matter, fixed by some soil inorganic components, and in mineral forms. The nutrients may also have to be in the form that is usable by the plant (available). Among these forms only nutrient present in the solution is readily available to plants and are replenished by the exchangeable and adsorbed forms of the nutrients. In most cases, only a small fraction of the nutrients present in soils are in available forms. A large portion of them is present either as part of organic matter, mineral or in the forms of sparingly soluble reaction products. The forms of all the essential nutrients in which they are most commonly used by the plant are given in Table 1.

Among these essential nutrients, some are available as positive ions and carry a positive (+) electrical charge while others are available as negative ions and carry a negative (-) electrical charge. Since majority of the soils have more negative charge than postive charge, cations are mostly adsorbed on the clay and organic matter. On the other hand most of the anions especially chloride and nitrates are found in the soil solution.

1.2 Function and Deficiency Symptoms of Nutrients

The each essential nutrient element is specific in its function in plant metabolism. The element may be the part

Table 1. Average concentration of nutrients in plants

Nutrient	Plant usable form	Average concentration in plant tissue
Hydrogen, H	H_2O	6.0%
Oxygen, O	H_2O and O_2	45.0%
Carbon, C	CO_2	45.0%
Nitrogen, N	NO_3^-, NH_4^+	1.5%
Potassium, K	K^+	1.0%
Calcium, Ca	Ca^{2+}	0.5%
Magnesium, Mg	Mg^{2+}	0.2%
Phosphorus, P	$H_2PO_4^-$, HPO_4^{2-}	0.2%
Sulphur, S	SO_4^{-2}	0.1%
Chlorine, Cl	Cl^-	100 mg/kg
Iron, Fe	Fe^{2+} and Fe^{3+}	100 mg/kg
Boron, B	H_3BO_3, $H_2BO_3^-$, HBO_3^{2-}, BO_3^-	20 mg/kg
Manganese Mn,	Mn^{2+}	20 mg/kg
Zinc, Zn	Zn^{2+}	20 mg/kg
Copper, Cu	Cu^{2+}	6 mg/kg
Molybdenum, Mo	MoO_4^{2-}	0.1 mg/kg

of structural component of plant's cells, cellular organizations, cellular metabolic processes, and energy transformations and in enzyme reactions. An inadequate or excess supply of one or more of these plant nutrients can cause metabolic disorders in plants. It may result a latent deficiency without visual symptoms or the appearance of visual deficiency or toxicity symptoms. Some of the best-known symptoms are a pale yellow-green colour for N deficiency, purplish spots or streaks for P deficiency and brown dead spots for K deficiency. Stunted growth is the most common deficiency symptom of all, but it is hard to identify unless there are more vigorous plants nearby for comparison. Important functions and deficiency symptoms of essential plant nutrients in brief are given in table 2.

Table 2. Functions and deficiency symptoms of plant nutrients

Nutrient	Functions in plants	Deficiency symptoms
Primary nutrients		
Nitrogen	Synthesis of amino acids, proteins, chlorophyll, nucleic acids, and coenzymes.	Light-green coloured leaves, lower leaves turn yellow and die in severe deficiency.
Phosphorous	Metabolic transfer processes, ATP, ADP, photosynthesis, respiration; and component of phospholipids.	Accumulation of carbohydrates results in dark-green colour leaves; restricted growth of plant tops and roots. Purplish leaves especially on the margins.
Potassium	Involved in sugars and starch formation; lipid metabolism and nitrogen fixation; neutralizes organic acids; transport of water and nutrients; disease and pest resistance; drought tolerance.	Marginal burning of leaves, curling of leaves.
Secondary nutrients		
Calcium	Component of cell walls; cell growth and cell division; co-factor for some enzymes.	Failure in the development of terminal bud, dead spot in the mid rib of some plants. In corn, tip of the new leaves may be covered with a sticky, gelatinous material that causes them to adhere to one another. Blossom end rot in tomatoes and Bitter pit in apples.

Contd...

Table 2. Contd...

Magnesium	Component of chlorophyll, hence essential for food synthesis in plant; promotes uptake of phosphorus and movement of sugars within the plant.	Light green leaves and yellowing of leaves similar to nitrogen. In rapeseed the leaves are cupped inward.
Sulphur	Essential constituent of sulphur containing amino acids, viz. cysteine, cystine and methionine; biological nitrogen fixation; involved in metabolic activities of vitamins, biotin, thiamine and coenzyme A; increase the oil quality in oil seed crops.	Fading of normal green colour of young leaves, followed by chlorosis. In Brassica, cupping and puckering of leaves inward.
Micronutrients		
Iron	Structural component of cytochromes, ferrochrome, and leghemoglobin and thus involved in oxidation-reduction reactions in respiration and photosynthesis.	Yellowing or whitening of young leaves. In severe deficiency in rice nurseries, or direct seeded rice or sorghum fields the entire plants may turn pale or white. Pale yellow interveinal chlorosis in stem.
Manganese	Photosynthesis-evolution of oxygen, oxidation-reduction processes, decarboxylation and hydrolysis reactions.	Interveinal discolouration-green veins against a pale background; whitening and abscission of leaves, gray speck of oats, marsh spots of peas, speckled yellow of sugar beet, blight of sugarcane.

Contd...

Table 2. Contd...

Zinc	Formation of auxins and chloroplasts; carbohydrate metabolism; stabilizing and structural orientation of membrane proteins.	Stunted growth, pale to white colouration of young leaves-white bud and white streaks in leaves of corn; brownish red (rusty) discolouration of leaves in rice known popularly as "Khaira" disease in rice. Corn, beans, citrus, and rice are indicator plants for zinc deficiency.
Copper	Constituent of chlorophyll, catalyst for respiration, carbohydrate, and protein metabolism.	Stunted growth, terminal leaf buds die, leaf tips become white, and leaves are narrowed and twisted.
Boron	Involved in germination and pollen tube growth; fruiting, cell division, nitrogen metabolism; facilitates ion uptake.	Terminal buds die, break-down of internal tissues in root crops (referred as brown or black heart), internal cork of apple, impairment of flowering and fruit development.
Molybdenum	Essential component of nitrate reductase and nitrogenase, thus important in nitrogen fixation by legumes; involved in protein biosynthesis; formation and viability of pollens and development of anthers.	Resemble nitrogen deficiency symptoms, whiptail disease of cauliflower; leaves show scorching and withering.
Chlorine	Involved in the evolution of oxygen in photosystem II of photosynthesis, rising of cell osmotic pressure.	Chlorotic leaves, some leaf necrosis.

The toxicity symptoms, induced by excess of plant nutrients and other mineral elements, are not specific for particular elements, they have many common characteristics. In most cases chlorosis and necrosis will develop starting in general at the leaf tips and/or margins of older leaves. Excessive large leaves with very enlarged cells and soft sponge tissue with or without chlorosis and necrosis of the leaf margins may also be symptom of N excess. Symptoms of "brown mottling of paddy rice" induced by excess iron, known as "Akagare" or "bronzing" in Asia, is characterized by small reddish brown spots or specks, which gradually extend on the lower dark green leaves. Sometimes the entire leaf changes brown and the lower leaves die off prematurely with dark green tint. The root becomes dark blue to black. Excess Al^{3+} in highly acidic soils (pH <5.0) is an important factor to limit growth of plant roots with short, stunted and retarded plant growth producing the foliage symptoms of phosphorus deficiency.

1.3 Soil Fertility Evaluation

Optimum productivity of any cropping system depends on the adequate supply of plant nutrients. In soil, crop uptake, leaching or erosion is constantly losing nutrients, while soil clays through the process of fixation can hold others. Although one or more nutrients are commonly applied to most crops, and released from organic matter and soil organisms, the quantity of nutrients removed in the harvested crop is generally much greater than the quantity added. This continuous loss will increase the nutrient related plant stress and yield and ultimately reduction in the productivity of the soil. When the soil does not supply sufficient nutrients for normal plant development and optimum productivity, application of supplemental nutrient is required. The proper rate of plant nutrient is determined by knowing the nutrient requirement of the crop and the nutrient supplying power of the soil. Hence, the evaluation of soil fertility becomes important. Soil fertility evaluation is essential for balanced nutrition of the crops. Balanced nutrients use refers to the application of essential plant nutrients in right amounts and proportions using correct methods and time of application suited for specific soil-crop-climatic situations. It helps in maintenance and improving soil productivity. Thus soil fertility evaluation is the key for adequate and balanced fertilization in crop production. Several techniques are commonly employed to assess the fertility status of the soils. A proper evaluation of the fertility of a soil before planting a crop helps in adopting appropriate measures to make up for the shortcomings and ensuring a good crop production. These diagnostic techniques are

1. Nutrient deficiency symptoms of plant,
2. Analysis of tissues from plant growing on the soil,
3. Biological tests in which the growth of either higher plants or certain micro-organisms is used as a measure of soil fertility, and
4. Soil testing.

1.3.1 Nutrient Deficiency Symptoms

As already mentioned, the plant requires sixteen essential nutrients for their optimum growth and development. When a plant badly need a certain nutrient element, it shows deficiency symptoms or hunger signs (Table 2). These symptoms are nutrient specific and show different patterns in crops for different essential nutrients. It is a good tool to detect deficiencies of nutrient in the field but these techniques has several limitations and are:

1. The visual symptoms may be caused by more than one nutrient.
2. Deficiency of one nutrient may be related to an excess quantity of another.

3. It is difficult to distinguish among the deficiency symptoms in the field, as disease or insect damage can resemble certain micronutrient deficiencies.
4. A visual symptom may be caused by more than one factor like low temperature, insect damage to roots, N deficiency, etc.
5. Nutrient deficiency symptoms are observed only after the crop has already suffered an irretrievable loss.

Deficiency symptoms are generally appear particularly in fast growing annuals due to their generous requirement of nutrients, where as slower-growing species on the same site do not show such symptoms. The vegetable are often first to show signs of deficiency because of rapid growth under intensive cultivations lead to high nutrient requirement and plant roots are unable to satisfy. In the perennials like trees, it may be slower initially, but they often become more severe year by year. There are some indicator plants, which show the nutrient deficiencies or excesses or respond more sensitively to such conditions as compared to others (Table 3).

Table 3. Indicator plants for nutrient deficiency/toxicity (Bergmann, 1992)

Plant	Nutrient deficiency/toxicity
Oat	Mg, Mn and Cu deficiencies
Wheat and barley	Mg, Cu and sometimes Mn deficiencies
Sugar and fodder beets	B and Mn deficiencies
Maize	N, P, K, Mg, Fe, Mn and Zn deficiencies
Potatoes	K, Mg and Mn deficiencies
Rape and fodder kale	N, P and Mg deficiencies
Brassica species	K and Mg deficiencies

Contd...

Table 3. Contd...

Sudan grass and lupins	Fe and Mn deficiencies
Celery and sunflower	B deficiency
Cauliflower	B and Mo deficiencies
Flax	Zn deficiency
Fruit trees	B, Cu, Fe, Mn and Zn deficiencies
Apple trees	K, Ca, Mg and B deficiencies
Berry bushes	Fe and Mn deficiencies
Barley	B, Mn and Al toxicities
Cucumber	N and P excess
Sugar beets	Cu excess

1.3.2 Biological Tests

It consists of raising a crop or a microbial culture in a field or in a sample of the soil, and estimating its fertility from the volume of crop or microbial count. It includes field trials, Neubaur seedlings technique, soil plaque, pot culture, etc. Although these methods are direct estimates of soil fertility, they are time consuming and, therefore, not well adapted to the practice of soil testing.

1.3.3 Plant Analysis

This technique involves two general types of plant analysis. One is the tissue test on the fresh tissue in the field, and other is a tissue analysis performed in the laboratory. In this technique, the amount of a given nutrient in plant is directly related to the availability of the nutrient in the soil. Tissue analysis is considered a more direct method of plant nutrient status evaluation than soil analysis, however, it involves a well-defined plant part analysis (Hallmark and Beverly, 1991). Among the several tissues to be considered for nutritional diagnosis purposes, leaves constitute the main plant sampling material

(Chapman and Brown, 1950). The improvement in tissue analysis techniques enabled the comparison of results from different soil type fertilization experiments. It is used more for fruit and vegetable crops, and also is becoming more useful for row crops and forage crops. The procedure for the analysis of essential nutrients in plants is given in this book.

1.3.4 Soil Testing

Soil testing is the chemical analysis that provides a guideline for lime and fertilizer needs of soils when considered in conjunction with post-fertilizer management and cropping history. Although plant analysis is extremely valuable in diagnosing nutrient stress, analysis of the soil is essential in determining the supplemental nutrient requirement of a crop. It refers to the chemical testing of soils for evaluating the fertility status and nutrient supplying power with respect to the application of fertilizers. It includes the analysis of available N, P, K, and micronutrients, texture, pH, $CaCO_3$ content and the parameters related to the amelioration of the chemically deterioted soils. At present, there are 497 soil testing laboratories in the country including 123 mobile soil testing units for undertaking on the spot soil testing and advising the farmers right at their fields. The primary advantage of soil testing when it is compared to the plant analysis is its ability to determine the nutrients status of the soil before the crop is planted.

The soil testing is done with following objectives:

- To optimize crop production.
- To protect the environment from contamination by runoff and leaching of excess fertilizers.
- To aid in the diagnosis of plant culture problems.
- To improve the soil's nutritional balance.

- To save money and conserve energy by applying only the amount of fertilizer needed.
- To identify soils contaminated with lead or other heavy metals.
- To prepare the soil fertility map of a village, subdivision, district, state, or country for the general status.

References

Arnon, D.J. and Stout, P.R. (1939). The essentiality of certain elements in minute quantity for plants with special reference to copper. *Plant Physiol.* 14:371-375.

Bergmann, W. (1992). *Nutritional Disorders of Plants-Development, Visual and Analytical Diagnosis*. Gustav Fischer Verlag Jena, Stuttgart, New York.

Chapman, H.D. and Brown, S.M. (1950). Analysis of orange leaves for diagnosing nutrient status with reference to potassium. *Hilgardia* 19:501-540.

Hallmark, W.B. and Beverly, R.B. (1991). Review-An update in the use of the Diagnosis and Recommendation Integrated System. *J. Fert.* 58:74-88.

2

Sampling of Soil for Nutrient Tests

Soil testing is a useful tool that can help to ensure the efficient use of applied plant nutrients. Soil tests provide a means for assessing the fertility status of a soil, but soil tests do not provide a direct measure of the actual quantity of plant available nutrients in the soil. Instead, soil tests measure the quantity of a nutrient element that is extractable from a soil by a particular chemical extracting solution. The measured quantity of extractable nutrient in a soil is then used to predict the crop yield response to application of the nutrient as fertilizer, manure, or other amendment. As soil test levels increase for a particular nutrient, the expected crop yield response to additions of that nutrient decreases. A soil-testing programme has four phases as follows:

1. Collection of soil samples,
2. Chemical analysis of soil samples,
3. Calibration and interpretation of the results of chemical analysis, and
4. Nutrient/Fertilizer Recommendation.

2.1 Collection of Soil Samples

The greatest challenge in soil sampling is obtaining a sample that reflects the true fertility of the field. A composite soil sample gives a mean analytical value representative of the soil sampling volume from which the composite sample is taken. Most fields are highly variable, especially when topography is rolling. However, even level fields that appear uniform can be highly variable. Certain general principles which should be strictly followed to make soil sampling correct is:

1. A series of cores taken according to some systematic grid layout of the area of equal diameter and comparable depth should be composited.
2. Separate soil cores should be analysed or replicate sets of composites made to determine statistical significance of results on the final composite.
3. The number of cores to be composited will depend on the variability of the soil, the degree of accuracy desired, the particular element or elements to be determined.
4. Cultivated soils are generally more variable than virgin soils, and acidic/saline are extremely variable.
5. Separate composite samples representing different segments of the soil profile or root zone should be taken.
6. Contamination from soil surface materials (crop residues, manures, fertilizers, etc.) should be avoided, also contamination of one soil depth.

Using a soil probe, soil auger, or spade can collect surface soil samples; however, for collecting subsoil samples, a spade is not very satisfactory. A post-hole digger can be used for collecting deep samples, but its use requires some special techniques.

The soil probe or tube is the most desirable tool for collecting soil samples. It will give a continuous core with minimal disturbance of the soil. The cores can be divided for the various depths. There should be very little contamination of subsoil sample with surface soil when using a soil probe. A soil probe cannot be used when the soil is too wet, too dry, or frozen. If the soil is frozen, the frozen layer will need to be fractured before a probe can be used. Soil probes cannot be used in soils that contain gravel. The soil auger can be used in soils that are frozen or contain gravel; however, great care must be taken to obtain representative samples and to avoid mixing of soil from different depths. The use of a soil auger in wet, sticky soils will result in mixing soil from different depths. A soil auger will not effectively gather dry, powdery soils. Use a soil auger only when a soil probe cannot be used.

2.1.1 Procedure

- Equipments required for the sampling are spade, auger, polythene bucket, and scale.
- Determine the soil unit (or plot).
- Make a traverse over the soil unit.
- Clean the site (with spade) from where soil sample is to be collected.
- Select the sites at random (20-30) in a zigzag (or criss-cross) manner (Fig.1) of cultural operations and natural trends of change such as slope.

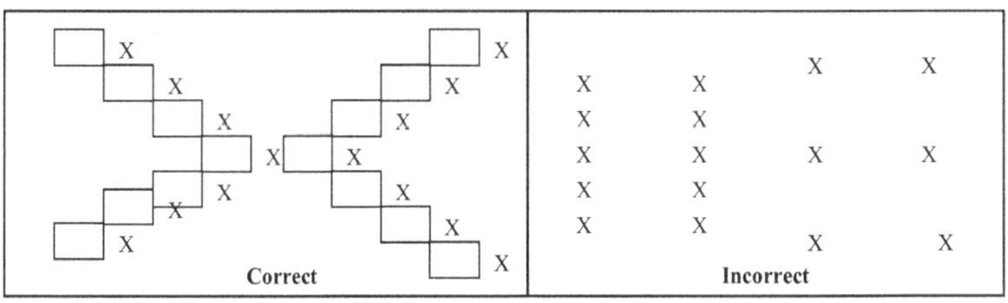

Fig. 1. A correct procedure for soil sampling

- Enough cores should be taken to represent the whole sampling area adequately.
- Dig soil with spade.
- Standing on opposite side, again insert the spade into soil.
- A lump of soil is removed.
- A pit of V shape is formed. Its depth should be 0-15 cm or 0-22.5 cm or 0-30 cm.
- Take out the soil- slice of ½ inch thick from both the exposed surface of the pit from top to bottom. To collect the soil-slice spade may be used.
- Collect the soil samples in a polythene bucket.
- The core should be of same volume and represent the same cross section of the sampling volume.
- There should be no chemical interactions of soil material composited that are significant to the objective.
- The soil unit selected for one composite sample should be homogeneous for the objective of the analysis.

The chemical analysis of the soil profiles is made for information on the chemical processes of soil development and long-range soil fertility (Jackson, 1973). The soil profiles to be sampled are selected:

- To represent agriculturally important soils.
- To represent soil development factors functionally.
- To represent the sequence of mineral weathering functionally.

Selecting soil types to represent the soil development and chemical weathering functionally simplifies the interpretation of results and provides the maximum of useful information per determination on the processes responsible for soil properties. A sequence of soil type is selected to isolate the effect of a single factor in so far as possible (such as soil types in sequence, climosequence, chronosequence, biosequence, lithosequence, catena, etc.).

2.1.2 Proper Sampling Depth

Surface (0-15cm depth) samples are frequently used for determining soil pH, lime need, organic matter, phosphorus, potassium, sulphur, and available micronutrients. Soil test correlations and calibrations for these tests are based on surface samples. Usually the tillage layer is considered to be the 0-15 cm or the 0-20 cm depth. For most soils, grain crop roots will penetrate to a depth of 1 m or more. Therefore, samples for nitrate-nitrogen ideally should be taken to a depth of 1 m, unless rooting depth is limited because of high water table, rock layer, coarse sand, or a gravel layer.

2.1.3 Time of Sampling

Winter is a good time for soil sampling except for testing for nitrate-nitrogen in sandy soils. However, soil samples can be taken any time excepting rainy season from the farmer's fields prvided there is sufficient time for the analysis of soil testing laboratories to give the soil test information to the farmers before the sowing of crops. Delaying sampling until spring provides for soil moisture replenishment from fall, winter and early spring precipitation, reflect the nitrate nitrogen distribution more accurately for when the crop is growing.

2.1.4 Drying

Moist, well-mixed samples are transferred to paper bags, cardboard boxes or aluminum trays of convenient size. The open sample container is then placed in a drying rack or cabinet equipped with exhaust fans to expedite air movement and moisture loss. This is especially critical for potassium analysis, which can be significantly influenced by drying temperatures. If nitrate analyses are involved, the soil should be dried or frozen within 12 hours of sampling. Spreading them out on a clean paper or cloth and blow-drying them with a fan can dry such samples. Where sample volume is not adequate to justify artificial drying, samples may be spread on clean surfaces, such as paper plates. Initial crushing of soil clods will decrease the time required for drying at room temperatures.

2.1.5 Crushing and Sieving

The nature of analyses to be conducted, plus presence of rocks or limestone concretions, determines initial steps to crushing. Crush samples designated for mechanical analyses with a wooden rolling pin after removing all stony material from the soil. Crush other samples with a flail-type grinder, a power-driven mortar and pestle, or some other crusher, which is designed to minimize contamination through carryover from one sample to another. If micronutrient analyses are to be performed, it is essential that all surfaces coming into contact with the soil be stainless steel, plastic or wooden, preferably in the order listed. Samples should be crushed until a major portion of the sample will pass 2 mm opening sieve. Crushing to pass a finer mesh sieve may be desirable for analysis utilizing less than one gram of soil.

2.2 Analysis of Soil Samples

For complete analysis, the following determinations are generally made:

A) Physical Constants

1. Apparent specific gravity or bulk density
2. True specific gravity or particle density
3. Pore space
4. Aggregate analysis
5. Moisture retention at 1/3 atmosphere and 15 atmosphere
6. Mechanical analysis of soils (% coarse sand, fine sand, silt and clay content)
7. Moisture (%)
8. Loss on ignition (%)
9. Calcium carbonate (%)

B) Chemical Analysis

1. pH
2. Electrical conductivity of saturation paste
3. Analysis of HCl extract or fusion with Na_2CO_3 for Al, Fe, Ca, Mg, P and K
4. Exchangeable cations such as Ca, Mg, K, and Na
5. Organic matter
6. Lime requirement
7. Total nitrogen
8. Available N, P, K
9. DTPA extractable cationic micronutrients (Fe, Mn, Cu & Zn)
10. Available boron and molybdenum

Most of the soil testing laboratories in our country is analyzing soil samples for pH, organic carbon (%), available N, P, and K. Now the emphasis is given for the analysis of available secondary nutrients, micronutrients and lime requirement of acid soils.

2.3 Calibration and Interpretation of the Results of Chemical Analysis

For the calibration of soil test data, a group of soils ranging in soil fertility from low to high in respect of the particular nutrient are selected and the test crop is grown on these soils with varying doses of that particular nutrient with basal dose of other nutrients. The most common method is to plot soil test values against the percentage yield and to calculate the relationship between soil test values and per cent yield response.

$$\% \text{ Yield} = (A-B) \times 100/A$$

Where A is the crop yield with adequate nutrient and B is the yield of control without the addition of the particular nutrient under study. The correlations obtained from the experiment under controlled conditions (green house) are given better as compared to field experiments because of the uncontrollable factors under the field conditions. However, from the large number of field experiments the relationships obtained can be accepted as a guide for the preparation of a fertilizer schedule. The lower the soil test value for a particular nutrient, the higher is the response to the added fertilizer nutrient. Assessment of nutrient requirement for different crops for different soils for fertilizer recommendation can be done by soil test crop response correlation studies.

Based on the contents of available nutrients, soil test values (N, P, K), can be grouped into classes of low, medium, and high. The ratings of nutrients used in India by the State Soil Testing Laboratories are given in table 4. The nutrient index (NI) of soils is calculated as follows:

$$NI = (L + 2M + 3H)/\text{Total number of soil samples tested}$$

Where, L, M, and H indicate the number of soil samples falling into low, medium, and high category of nutrient availability. The value of NI arbitrarily chosen below 1.5, 1.5-2.5, and above 2.5 as an index of nutrient status indicate

Table 4. Rating chart of soil test values (Muhr *et al.*, 1965)

Nutrient	Category of nutrient availability		
Organic carbon (%)	<0.50	0.5-0.75	>0.75
Available P (kg/ha)	<10	10-25	>25
Available potassium (kg/ha)	<120	120-280	>280

the low, medium and high category of nutrient availability, respectively. The different categories of soil fertility indices are most commonly interpreted on the basis of the probability of an economic response to adding additional nutrients are given in table 5. Soil testing for every three years is recommended with this approach.

Table 5. Soil test interpretation for different categories of nutrient availability

Category	Definition and interpretation
Below optimum	• Indicates that the nutrient is probably deficient and that the deficiency will likely limit crop growth. • High probability of a profitable return from correcting a low level. • Recommendations for a soil testing "below optimum" are increasing the soil in test level above optimum and to maintain it that level.
Optimum	• Indicates that the nutrient is probably adequate and will likely not limit growth in a typical growing season. • There is a low probability of a profitable return from increasing the soil test level above optimum. • Recommendations for a soil testing "optimum" are designed to offset crop removal in order to maintain the nutrient in the optimum range.

Contd...

Table 5. Contd...

Above optimum	• If testing is done on an annual basis, no maintenance fertiliser is needed when soil tests in the optimum range. • Indicates that the nutrient is more than adequate and will not limit crop growth. • Very low probability of a profitable return from applying additional nutrients to a soil testing "above optimum". • No fertilizer is recommended on these soils. Too much of a plant nutrient may cause a nutrient imbalance in the soil and, as a result, in the plant, which may adversely affect plant growth and environmental quality.

It is often difficult to clearly distinguish whether the critical level is a sufficiency level for a crop fertilization or an optimum level for soil fertilization because of inherent variability in soil test calibration data and the high level of uncertainty in determining either the actual sufficiency level for a crop or the optimum value for a soil.

2.4 Nutrient/Fertilizer Recommendation

The main objective of the soil testing is to recommend the fertilizer for efficient and economic crop production. Soil tests give the relative level of a nutrient in the soil and must be calibrated to relate its level to the likely response to application of fertilizer. Output for different crops for different doses of fertilizers are obtained from the response curves. On the basis of costs of inputs and outputs, the maximum yield, maximum profit/ha, maximum rate of return/rupee of investment, and minimum profitable application information are derived from the developed soil test crop response correlation equations. The quantity of a nutrient taken up by the crop to produce the target yield can be estimated. The following equation is used to solve the needed fertilizer rate to get the targeted yield.

$$X = (Y-Z)/R$$

Where X is the amount of the element to be supplied by fertilizer,

R is the proportion of the element applied recovered by the crop,

Y is the estimated amount taken by the crop,

Z is the estimated amount supplied by the soil.

On the basis of equation developed, a farmer can make his choice accordingly on the basis of soil testing of his field under operation. However, the amount supplied by soil is influenced by the amount returned in crop residues. The "fertilize the soil" approach is most appropriate for long-term management where a return from the investment in building soil test nutrient values into the optimum range will be achieved.

References

Muhr, G.R., Datta, N.P., Sankrabamoney, H., Leley, V.K. and Donahue, R.L. (1965). *Soil Testing in India*. USAID Mission to India, New Delhi.

Jackson, M.L. (1973), *Soil Chemical Analysis* (Indian edition). Prentice Hall of India Pvt. Ltd. New Delhi.

3

Soil Analysis

3.1 Important Soil Physical Properties

3.1.1 Analysis of Soil Texture

Soil texture refers to the relative proportions of sand, silt, and clay on weight basis. Soil texture was recognized from very early times by experienced farmers from the feel of moist soil placed between the thumb and forefinger. The size groups arbitrarily classified by the International Society of Soil Science (ISSS) into clay (< 0.002 mm), silt (0.002-0.02 mm), fine sand (0.02-0.2 mm), and coarse sand (0.2-2.0 mm). Each soil textural name specifies the weight percent of each fraction falls within certain defined limits. The determination of the percentage of the soil separate present in a sample is called particle-size analysis.

Each type of particle present makes its contribution to the nature of the soil as a whole. Sandy soils are usually quite permeable to air, water, and roots, but they have two important limitations. One is their relatively low water-holding capacity; the second is that they are poor storehouses for plant nutrients. These limitations of sandy soils can, of course, be overcome if both fertilizer and irrigation water are available, but the costs are large.

Clay not only have a large surface area per gram, it also has electrically charged particles. These charges give clay the capacity to attract plant nutrient ions to its surface in forms available to plants, but sand lacks this capacity. Clay holds much more water than sands because they have a large surface area to be coated with water. One important problem with clay soils is their stickiness. When wet, they stick to ploughs, boots, or whatever they contact. Loam and silt loam soils are highly desirable for most uses. They have enough clay to store adequate amount of water and plant nutrients for optimum plant growth but not so much clay as to cause poor aeration or to make working with them difficult.

3.1.1.1 Determination of soil texture by hydrometer method (Buoyoucous, 1927)

The percentage of silt and clay are determined by treating a weighed soil sample with a sodium compound (usually sodium polyphosphate or sodium oxalate) and dispersing it by means of a shaker or an electric mixture. Dispersion breaks the soil aggregates apart so the particles act individually in the analysis. The suspension of dispersed soil is transferred to a tall cylinder and shaken end over end a few times, the place upright. The particles fall through the water, and sand soon settles from the upper few centimeters, leaving behind silt and clay. The percentage of sand, silt, and clay can be determined from the data obtained in the proceeding analyses.

3.1.1.1.1 Apparatus

- Buoyoucous soil Hydrometer
- Mechanical Shaker
- Thermometer
- 1 liter measuring cylinder
- Stirrer/Plunger

3.1.1.1.2 Reagent

1. 5% sodium hexametaphosphate: w/v aqueous
2. 30% H_2O_2-reagent grade
3. 2 N HCl: 16.7 ml conc. HCl dissolved in 100 ml volumetric flask in water.
4. 1 N NH_4OH

3.1.1.1.3 Procedure

1. Take 50 g of soil sample in a 500 ml conical flask.
2. Treat the soil with H_2O_2 to destroy the organic matter.
3. Add 200 ml of distilled water and 100 ml of Sodium hexametaphosphate solution and mix well with stirrer and keep the suspension for 4-5 hours or overnight.
4. Shake the suspension vigorously several times on a mechanical shaker.
5. Transfer the dispersed soil medium into a 1 liter measuring cylinder and make the volume to 1 litre.
6. Place a rubber stopper over the mouth of the cylinder, and shake it vigorously, back and forth, for 1 minute.
7. Place the cylinder on a table and note the time immediately.
8. Immediately place the hydrometer in the suspension.
9. Record the reading exactly 4 minutes after placement of hydrometer.
10. Reinsert the hydrometer at the end of 2 hours and note the hydrometer reading.

Note: If Temp >67^0 F then hydrometer reading has to be added 0.2 per degree and if <67^oF then reading has to be subtracted 0.2 per degree.

3.1.1.1.4 Calculation

A. % Silt + clay = value of 1st reading X 2

B. % Clay = value of 2nd reading X 2

C. % Silt = A - B

D. % Sand = 100 - (Silt + Clay)

3.1.1.1.5 Estimation of soil textural classes

The textural class of the soil corresponding to specified percentage of sand, silt and clay can be determined from the triangle shown in Figure 2. The triangle is divided into 12 areas containing all possible proportions of sand, silt and clay. The numbers on the three scales are angled to indicate the slopes of the lines to which they apply. The intersection of lines in the figure for sandy clay loam represents 30 % clay (along a horizontal line), 10 % silt (along a line parallel to the left side of the triangle), and 60% sand (along a line parallel to the right side of the triangle). Line may be similarly traced on the triangle to show the soil textural classes having different percentage of sand, silt and clay.

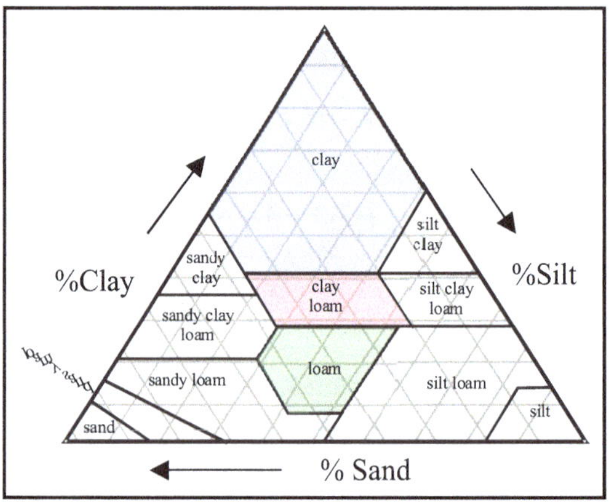

Fig. 2. Textural triangle showing the centers of sand, silt and clay (%) in the major soil textural classes

Sometimes, the point representing the texture of a soil sample happens to fall exactly on the line between two textural names. It is customary to use the name of the finer fraction when this happens. For example, a sample containing 40 % clay, 30 % silt and 30 % sand is called clay rather than clay loam.

3.1.1.1.6 Texture by feel

It is often necessary to determine soil texture quickly or in some laboratories where determination is not possible. The soil analysis workers must learn to estimate texture by feeling the soil between their thumb and fingers. The plasticity of soil is probably the best guide to estimate the clay content. It is evaluated by attempting to squeeze the moist soil into a thin ribbon as it is pressed with a rolling motion between thumb and forefinger. The method for estimating soil texture (Portland Cement Association, Illinois, USA, 1962) is given below.

Place approximately 25 g of soil in the palm; add water drop-by-drop and kneed to break down structure. Soil is at the proper consistency when plastic and moldable like moist putty. Place the soil ball between thumb and forefinger and push the soil to form a ribbon and its extension over the forefinger until it breaks from its own weight and observe:

Observation	Texture
No formation of ball	Sand
No formation of ribbon	Sandy loam
If the ribbon is < 2.5 cm long and break and the thoroughly-wetted sample feels:	
1. Gritty	Sandy loam
2. Smooth	Silt loam
3. Neither gritty or smooth	Loam

Contd...

If the ribbon is between 2.5-5.0 cm long and break and the thoroughly -wetted sample feels:	
1. Gritty 2. Smooth 3. Neither gritty or smooth	Sandy clay loam Silty clay loam Clay loam
If the ribbon is between *5.0 cm* long more than and the thoroughly-wetted sample feels:	
1. Gritty 2. Smooth 3. Neither gritty or smooth	Sandy clay Silty clay Clay

A practiced hand can identify by feel any of the texture shown above.

3.1.2 Bulk Density

The bulk density is the weight of the soil solids per unit volume of total soil. The bulk density of the A horizons of mineral soils is usually between 1.0 and 1.6 g/cm^3. The variation occurs mostly from the differences in total pore space. As a general rule, the heavy-textured soils have more pore space and lower bulk densities than sandy soils. Organic matter decreases bulk density in two ways: (1) organic matter is much lighter in weight than a corresponding volume of mineral matter, and (2) organic matter increases the aggregate stability of soils. The soil bulk density tends to increase when cultivation causes a loss of organic matter from the soil. The bulk density is useful to calculate the weight of various masses of soils.

3.1.2.1 Equipment and Apparatus

1. Core sampler of desire dimension 2. Vernier caliper 3. Can boxes 4. Oven 5. Balance 6. Knife spatula 7. Square metal plate with a rod at the center.

3.1.2.2 Procedure

1. Place the core sampler vertically at the selected site and push the core sampler with metal plate rod uniformly into the soil deep enough to fill the sampler can.
2. Dig out the sampler by means of a khurpi or spade and remove the sample can without disturbing the soil core in the can.
3. Trim off the extra soil from both ends of the sample can with a sharp knife.
4. Weigh the sample can along with the soil up to 1 g of accuracy.
5. Dry the core soil with metal cylinder in an oven at 105°C, cool and take the weight of metal cylinder + soil.
6. Remove the soil core from the metal cylinder and weigh the metal cylinder.
7. Subtract the mass of metal cylinder from the mass of over-dried soil core + metal cylinder to obtain the mass of oven dry soil.
8. Measure the radius (r) and height (h) of the metal cylinder in cm by the Vernier Calipre and calculate the volume of soil core by the equation of $\pi 2r^2h$ in cm^3.
9. Bulk density = mass of oven dry soil core (g)/volume of soil core (cm^3) = g/ cm^3

3.1.3 Particle Density

Particle density is the average density of the soil particles. It is expressed g/cm^3. The particle densities of mineral soils vary much less as compared to bulk densities. The usual practice is to assume that the particle density of a mineral soil is 2.65 g/cm^3 unless very precise data are sought. Particle density is not altered by differences in particle size and not by changes in pore space.

3.1.3.1 Estimation

1. Take a clean dry 100 ml capacity graduated cylinder.
2. Add distilled water to fill the graduated cylinder up to 50 ml mark.
3. Weigh 40 g of over-dry soil and transfer it to the same cylinder filled with 50 ml of water.
4. Stir the contents thoroughly for 10 minutes with a glass rod and allow the particles to settle.
5. Note the increase in the volume of water on adding 40 g of soil to measure the volume of water displaced by the soil in cm^3.
6. Particle density (g/cm^3) = mass of oven-dry soil/ volume of water displaced by the soil (cm^3)

3.1.4 Percent Pore Space

The percent pore space in soils is calculated from the data on bulk density and particle density. On the basis of these two values, the percent of the soil volume occupied by solids is

% Solid = bulk density/particle density X 100

and

% Pore space = 100 - % solid

If preferred, these two can be combined into one:

% Pore space = 100 – (bulk density/particle density) X 100

or usually

% Pore space = [1 – (bulk density/2.65)] X 100

It should be noted that increasing bulk density signifies an increase in percent solid and a corresponding decrease in percent pore space.

Reference

Bouyoucous, G. J. (1927). The hydrometer as a new method for the mechanical analysis of soils. *Soil Sci.* *23*:343-353.

3.2 Soil pH

Soil pH is a measure of hydronium ion (H_3O^+ or, commonly, H^+) activity in a soil suspension. Soil pH is defined as the negative logarithm (base 10) of the H^+ activity (moles per liter) in the soil solution. As the activity of H^+ in the soil solution increases, the soil pH value decreases. Soils with pH values below 7 are referred to as "acid"; pH values above 7 are referred to as "alkaline"; soils at pH 7 are referred to as "neutral." In most soils, the soil pH is buffered by several components of the solid phase, including hydroxyl aluminum monomers and polymers, soil organic matter, and undissolved carbonates in soils. This property influences many aspects of crop production and soil chemistry, including availability of nutrients and toxic substances, activity and diversity of microbial populations, and activity of certain pesticides.

3.2.1 Principle

A pH meter is an actually a direct current amplifier. It measures the *emf*, which appears across the electrodes upon being immersed in solution, soil suspension or irrigation water. The glass and calomel electrodes are immersed in the test solution and the *emf* determined by the electron tube voltmeter. The meters graduated to read directly in pH unit along with the *emf* (milli-volt) scale. A standard buffer solution (of known pH) is used to calibrate the instrument before determining the pH of test solutions.

3.2.2 Apparatus

1. pH meter
2. Top loading balance

3.2.3 Reagents

1. Buffer solution of pH 4.0, 7.0 and 9.2: Buffer tablets of this pH are available from the chemical manufacturers. Dissolve one tablet of pH 7.0 in pure distilled water and make the volume to 100 ml, shake well. Likewise, prepare buffer solution of pH 4.0 and 9.2.

3.2.4 Procedure

Weigh 20 g of soil sample in 100 ml beaker and add 40 ml of distilled water (Soil: water; 1:2)

↓ Stirring

Stir with glass rod occasionally for about 30 minutes or shake for 5 minutes with a reciprocating mechanical shaker

↓ Standardizing the pH meter

Immerse the glass electrode in buffer solution (say pH 7.0) and adjust the pH reading to pH 7.0 of the pH meter

↓

Wash the electrode with distilled water and repeat the same procedure for another known standard buffer solution (say pH 4.0 for acid soil or pH 9.2 for alkaline soil)

↓ Sample reading

Immerse the electrodes of the pH meter in the soil-water suspension and read the pH

3.2.5 Interpretations

S.No	Acidity class	pH range
1	Extremely acidic	<4.5
2	Very strongly acidic	4.6-5.0
3	Strongly acidic	5.1-5.5
4	Medium acidic	5.6-6.0
5	Slightly acidic	6.1-6.5
6	Neutral	6.6-7.3
7	Mildly alkaline	7.4-8.0
8	Strongly alkaline	8.1-9.0
9	Very strongly alkaline	>9.0

3.3 Exchangeable Acidity and Lime Requirment of Acid Soil

Soil acidification is a natural process in soils from both acidic and basic parent materials. Soils usually become more acidic with time due to weathering. The process often exacerbated by human activities, high rainfall in humid areas, intensive use of acid fertilizers and acid rain. As a result, soil acidity continues to be a problem and is increasingly becoming a limiting factors for agricultural production in humid high rainfall areas of the world. Excessive soil acidity must be neutralized in order to attain a desired soil pH and soil environment for better nutrient availability.

3.3.1 Exchangeable Acidity

The soil pH measures the intensity of hydrogen ions concentration (activity) in soil solution and does not take into account the activities of other ions such as Al, Fe, and Mn, which can influence the acidity of soil. The titrateable electrovalent and intermediate between electrovalent and covalent titrateable hydrogen ions are exchangeable and hydrolytic forms is produced by hydrolysis of aluminium and iron salts.

The conductometric titration of acid soils can be used to estimate the proportion of titrateable hydrogen to titrateable aluminium. The conventional method of analysis employ solution of neutral salt to leach the soil to displace the exchangeable hydrogen and aluminium ions held by exchangeable acidity. Yuan (1959) recommended the use of 1N KCl solution as extractant with the subsequent addition of NaF or KF to the leachate to eliminate the effect of aluminium by formation of sodium-fluro-aluminate.

$$AlCl_3 + 6NaF = Na_3AlF_6 + 3NaCl$$

This complex is neutral to phenolphthalein indicator and treated extractant can be titrated with standard solution of NaOH using that indicator to give a measure of exchangeable hydrogen. Thus, this extractant is used for the quantitative estimation of entire acidity (exchangeable hydrogen and aluminium).

3.3.1.1 Apparatus

Shaker, 250 ml conical flasks, funnels, filter, burette, and 50 and 25 ml pipettes.

3.3.1.2 Reagent

1N KCL solution: Dissolve 74.56 g of KCl/litre in water.

KF solution: Dissolve 58.1 g of KF/litre of water and titrate with NaOH to a end point using phenolphthalein indicator.

0.1N HCl standard.

0.1N NaOH standard.

Phenolphthalein indicator: Dissolve 1.0 g of indicator in 100 ml ethyl alcohol.

3.3.1.3 Extraction and estimation of exchangeable acidity

1. Take 10 g of air-dry soil in 250 ml conical flask.
2. Add 50 ml of 1N KCl solution and mix.
3. Shake the content for 30 minutes on a shaker.
4. Transfer the suspension to a funnel fitted with filter paper mounted on a 250 ml conical flask.
5. Follow with an additional volume of 100 ml 1N KCl in 25 ml increments to a total of 150 ml.
6. Add 4 or 5 drop of phenolphthalein indicator in the leachate and titrate with 0.1N NaOH to the first permanent pink endpoint. Also similarly titrate the 150 ml KCl solution for blank.
7. Record the value of titre for the estimation of exchangeable H^+ + Al^{3+} ions acidity.
8. Add 10 ml of 1N KF to the previously 0.1N NaOH titrated extract and titrate with 0.1N HCl until pink colour disappears.
9. Wait for 30 minutes and add additional HCl to a clear end point.

3.3.1.4 Calculation

me acidity = (ml 0.1N NaOH sample - 0.1N NaOH blank) X 100/weigth of soil sample

me exchangeable Al^{3+} = ml 0.1N HCl X 100/weight of soil sample

me H^+ = me acidity –me exchangeable Al^{3+}

3.3.2 Lime Requirement

Lime requirement (LR) is the amount of lime needed to raise pH to a selected level or to reduce the content of toxic substance to a certain level. Lime requirement methods consisting of buffered solutions merely determine a proportion of the exchange acidity of a soil. The amount of

$CaCO_3$, required to achieve a soil pH is then obtained by calibrating the pH of the soil-buffer mixtures against the LR established by titrating soils with $Ca(OH)_2$, or incubating them with various amounts of $CaCO_3$. As buffered tests are a rapid and convenient with titrating or incubating soils, several methods have been proposed over the years. However, none of them is universally accepted for use under all circumstances. Shoemaker *et al.* (1961) original single-buffer method also known as SMP method is used with soils containing appreciable amounts of exchangeable aluminium and is expected to have LR values greater than 2 t/ha. This method has found increasing favour due to its being rapid test method and most commonly used as compared to other procedures. So, this method is given here.

3.3.2.1 Apparatus

1. pH meter
2. Electric shaker
3. Beaker, etc.

3.3.2.2 Reagents

Buffer solution: Dissolve 2.0 g of calcium acetate, 40 g of calcium chloride, 3.0 g of potassium chromate, 1.2 g of P-nitrophenol and 2.0 ml of triethanolamine in 800 ml of water. Adjust the pH to 7.5 using dilute HCl or NaOH solution, and dilute to 1 litre.

3.3.2.3 Procedure

Take 10 g of soil in a beaker and add 20 ml of buffer solution

↓

Shake for 10 minutes

↓

Determine the pH by glass electrode pH meter

↓

The depression in pH of buffer is proportional to the lime requirement, which can be known from the **table 6**

Table 6. Lime requirement to adjust soil pH to a desired level based on pH SMP

Soil Buffer pH	Lime ($CaCO_3$) requirement (t/ha)	Soil Buffer pH	Lime ($CaCO_3$) requirement (t/ha)
6.7	3.95	5.7	18.77
6.6	5.43	5.6	20.25
6.5	6.92	5.5	21.24
6.4	8.40	5.4	22.47
6.3	9.88	5.3	24.95
6.2	11.12	5.2	27.17
6.1	12.84	5.1	28.90
6.0	14.37	5.0	30.63
5.9	15.81	4.9	33.09
5.8	17.30	4.8	32.36

References

Shoemaker, H.E., McLean, E.O. and Pratt, P.F. (1961). Buffer methods for determination of lime requirement of soils with appreciable amount of exchangeable aluminium. *Soil Sci. Soc. Am. Proc.* 24:274-277.

Yuan, T L. (1959). Determination of exchangeable hydrogen in soils by titration method. *Soil Sci.* 88:164-167.

3.4 Soluble Salts

The soluble salts present in soils include the dissolved inorganic soluble salts. Of the soluble anions, sulphates and chlorides predominate, but exchangeable cations in equilibrium with carbonic acid (H_2CO_3) dissolved in the soil, yielding soluble carbonates (CO_3^{-2}) and bicarbonates (HCO_3^-) of the metallic cations and leaving the corresponding amount of hydrogen ion on the exchange complex. Nitrates, phosphates, and silicates occur in varying concentrations, often negligible but sometimes may occur in excess. The cations involved are usually calcium, magnesium, potassium, and sodium. In acid soils, aluminium, iron, and manganese cations become more important in humid tropical high rainfall areas. When a soil contains an excess of soluble salts, it is termed a *saline* soil. It is also called a *"white alkali"* soil as a result white saline crust that sometimes appears on drying of soil. The management of saline soils is evaluated from the measurements of such soluble salts. If the sodium content increases the soil becomes saline-alkali soil, the pH remaining below 8.5.

The presence of enough salts in soluble form causes interference with normal seed germination, plant growth, or plant intake of water. Plants vary greatly in their tolerance to salinity, and injury due to salts is dependent partly upon soil conditions other than salt content. The harmful effects of excess salts arise from Na^+, Cl^-, SO_4^{2-} but seldom from Ca^{2+}, Mg^{2+}, or CO_3^{-2}. Chlorosis is sometimes associated with high HCO_3^- concentration coupled with high pH. Toxicity of boron to plants even in small amounts is important in irrigation waters and saline soils. A very broad generalization of plant susceptibility to soluble salts is given in the USDA Salinity Handbook, which is based upon the conductivity of soil saturation extracts.

Conductivity (mmhos/cm)	
0-2	Salinity effects negligible
2-4	Very sensitive crops affected (citrus, beans)
4-8	Many crops affected
8-16	Only salt-tolerant crops yield satisfactorily (wheat, grapes, olives)
>16	Only very sensitive salt-tolerant crops yield satisfactorily (dates, barley, sugar-beet)

3.4.1 Extraction of Soil Water-Soluble Salts

Soluble salts in soils can be determined from the measurements made (i) on aqueous extracts of soil samples, (ii) on samples of soil water itself obtained from the soil, (iii) in soil using buried porous salinity sensors that imbibe and equilibrate with soil water, and (iv) in soil or soil paste using four-electrode probes or electromagnetic (EM) systems (Rhoades, 1982). The appropriate method of measuring soil salinity must be selected for the specific condition and purpose. Soil salinity is conveniently defined and measured on aqueous extracts of saturated soil pastes (US Salinity Laboratory Staff, 1954). The saturated soil is used because it is the lowest reproducible ratio for which extract for analysis can be readily removed from the soil with pressure or vacuum and often related to field soil water contents. Other extraction ratios, 1:1, 1:5, etc., are easier to use as compared to saturation. However, they are not so well related to field soil water contents due to peptization, hydrolysis, cation exchange, and mineral dissolution. When relative changes rather than absolute solute concentration are monitored, these wider extraction ratios may be used to advantage.

3.4.1.1 Saturation extract of soil

The saturated paste of soil is made by manually mixing soil with water until certain specified conditions are satisfied. Weigh 200-400 g of air-dry soil in a suitable container and add distilled water while stirring with a spatula until it is nearly saturated. At saturation the soil surface glistens, the soil flows slightly if the container is tipped, slides freely and cleanly off a spatula, and consolidate easily by side of the spatula. Allow the paste to stand for four hours (preferably overnight) and repeat the saturation tests; there should be no free water on the paste surface. If too much water has been added a little drier soil can be mixed. Record the increase in weight, which is the amount of water added. Calculate the saturation water percentage from the weight of oven-dry soil and the sum of the weights of water added and the initially present in the air-dry soil. After allowing the saturated soil paste to stand 4 or more hours, transfer it to a Buchner filter funnel fitted with highly retentive filter paper. Connect the tube of vacuum pressure tube to the filtrate receiving clean dry 1 litre conical flask, on which mounted buchner funnel. Apply vacuum, and collect the filtrate in the conical flask by suction. If the initial filtrate is turbid, refilter or discard it. Terminate the filtration when air passes through the filter. Colouration of the extract by dissolved organic matter does not appreciably affect the conductance or chemical analysis. Add one-drop of 0.1% sodium hexametaphosphate solution for each 25 ml of extract (Rhoades, 1982).

3.4.1.2 Extraction with pressure membrane

The soil solution can be obtained by pressure membrane. Extraction of soil solution by means of the pressure membrane is slower and requires a large soil samples as compared to water extracts, but has the advantage that the soil solution is obtained without dissolution of additional materials from the soil.

3.4.1.3 1:1 and 1:5 soils: water extracts

Weigh 100-200 g of dry soil into the 500 ml conical flask. Add distilled water to the required amounts either 1:1, 1:2 or 1:5 soil: water ratios. Stopper the flask and shake for a period of one hour in a rotary shaker. Filter the suspension on a Buchner filter funnel fitted with highly retentive filter paper. Add one-drop of 0.1% sodium hexametaphosphate solution for each 25 ml of extract.

3.4.2 Analysis of Soluble Salts in Water Extracts

The analysis of water soluble salts in the soils include electrical conductivity, pH, calcium, magnesium, sodium, sulphate, chloride, boron, nitrate, carbonate, bicarbonate, sodium adsorption ratios and residual carbonate. Follow the method of analysis of these constituents as described for the water quality analysis (5.2.1-5.2.9).

% Salt in soil = 0.064 X conductance (dSm^{-1}) X % water in soil at extraction/100

The factor (0.064 X % water in soil at extraction) becomes 0.320 in 1:5 extract, (Jackson, 1973).

References

Jackson, M.L. (1973). *Soil Chemical Analysis* (Indian edition). Prentice Hall of India Pvt. Ltd. New Delhi.

Rhoades, J.D. (1982). Soluble salts. In: *Methods of Soil Analysis*, Part 2 (eds. Page, A.L., Miller, R.H. and Keeney, D.R.), pp. 167-179. ASA and SSSA, Madison, Wisconsin, U.S.A.

U. S. Salinity Laboratory Staff (1954), L.A. Richards Led. Diagnosis and Improvement of Saline and Alkali Soils. U. S. Deptt. of Hand Book No. 60.

3.5 Organic Carbon in Soil (Walkley and Black's rapid titration method, 1934)

Soil organic matter consists essentially of a series of products that ranges from decayed plant and animal tissues to fairly amorphous brown to black material, defined as soil humus. In addition to this, soil organic matter contains living and dead microbial tissues, microbially synthesized compounds and derivatives of these materials, produced as a result of microbial decay. The importance of soil organic matter in supplying nutrients, contributing CEC, improving soil structure, etc., is well recognised. A major portion of N (95 to 99% of the total), P (33-67% of the total), and S (75% of the total) in soils occurs in organic combinations, which mineralize to release the nutrients in inorganic forms to be used by plants. The organic matter content of a soil varies from <0.10% in very sandy arid soils to more than 80% in peats and mucks. Soil organic matter contains 5% N, and 0.5% each of P and S, there by having an N: P: S ratio of 10:1:1. The organic matter content of a soil is estimated by the amount of carbon present, on an average, is 52-58% of the organic matter, and the C: N ratio is 10-15:1. In order to find out the amount of humified organic material, the organic carbon content is multiplied by 1.724, which represents the ratio between the humus and the organic carbon (100/58), and is known as the 'van Bemmlen factor'.

3.5.1 Principle

Soil organic carbon is determined by dry combustion and wet digestion methods. The dry combustion method gives absolute values and is useful for very accurate work. But for the ordinary routine work, Walkley and Black's (1934 and 1947) rapid titration method is useful. In this method, the soil is digested with chromic acid and sulphuric acid making use of the heat of dilution of sulphuric acid and the soil organic carbon is, thus, oxidised to CO_2. The highest temperature attained by the heat of dilution of sulphuric acid is approximately 120°C, which is sufficient to oxidize the active forms of the soil organic carbon, but not the mere inert form of carbon in soil. The excess of chromic acid, not reduced by the organic matter of the soil, is determined by titration with standard ferrous sulphate or ferrous ammonium sulphate solution, in the presence of NaF or phosphoric acid and diphenylamine indicator.

3.5.2 Apparatus

1. Conical flasks (500ml)
2. Burette, etc.

3.5.3 Reagents

1. 1 N Potassium dichromate ($K_2Cr_2O_7$): Dissolve 49.09 g of AR grade $K_2Cr_2O_7$ in distilled water and make the volume to 1 litre.
2. 0.5 N Ferrous ammonium sulphate [$Fe(NH_4)_2(SO_4)_2 \cdot 6H_2O$]: Dissolve 186 g of $Fe(NH_4)_2(SO_4)_2 \cdot 6H_2O$ in

distilled water. Add 7.5 ml conc. sulphuric acid and make up the volume to 1 litre with distilled water.

3. Conc. sulphuric acid not less 96% (H_2SO_4): containing 1.25% silver sulphate in case of soils containing chloride. (add 1 g of Ag_2SO_4 in every 100 ml of conc. H_2SO_4).
4. Diphenylamine indicator: Dissolve 0.5 g of diphenylamine in a mixture of 20 ml of water and 100 ml of conc. H_2SO_4.
5. Ortho phosphoric acid (H_3PO_4) (85%) and or sodium fluoride (pure)

3.5.4 Procedure

Weigh 0.5 g soil in a 500 ml conical flask

↓

Add 10 ml of 1 N $K_2Cr_2O_7$

↓

Add 20 ml of conc. H_2SO_4 through sides of the flask

↓

Shake gently to mix and stand for 30 minutes

↓

Add 200 ml distilled water and 10 ml of H_3PO_4 or 0.5 g of NaF

↓

Add 1 ml of diphenylamine indicator

↓

Titrate with 0.5N $Fe(NH_4)_2(SO_4)_2.6H_2O_4$ (FAS)

↓

End point: Colour changes from *violet* through *blue to bright green*

↓

Run a blank similar to the soil sample procedure without soil.

3.5.5 Calculation

% Organic carbon in soil = N X (B – T) X 0.003 X 100/ W.S

Where,

N = Normality of FAS

B = Blank burette reading (ml)

T = Sample burette reading (ml)

W.S = Weight of soil sample (g)

% Organic Matter in the soil = % Organic Carbon X 1.724

References

Walkley, A., and I.A. Black. (1934). An examination of Degtjareff method for determining soil organic matter and a proposed modification of the chromic acid titration method. *Soil Sci.* 37: 29-37.

Walkley, A. (1947). A critical examination of a rapid method for determining organic carbon in soils: Effect of variations in digestion conditions and of inorganic soil constituents. *Soil Sci.* 63: 251-264.

3.6 Available Nitrogen (Alkaline permanganate method)

This method was originally developed for estimating the easily oxidizable organic nitrogen fraction in organic manures and fertilizer (AOAC, 1930). Its adoption for soil testing in India, method developed by Subbiah and Asija, (1956) based on fairly good relationships with responses of paddy-rice and wheat in the field and pot cultures (Tamhane and Subbiah, 1962). The alkaline permanganate releases N from the soil organic N pool by processes of oxidation and hydrolysis (Stanford, 1978). This method estimates only the ammonium N released from the soil organic matter by alkaline permanganate in addition to ammonium-N already present in the soil. It does not include the nitrate-N present in the soil, therefore results poorer relationships for upland crops (Sahrawat and Burford, 1982). Sahrawat and Burford (1982) recommended distillation of ammonium -N from alkaline permanganate flask then addition of Devarda's alloy followed by further distillation to include nitrite and nitrate-N present in the soil for the estimation of available N.

3.6.1 Principle

A known weight of the soil is mixed with excess of alkaline $KMnO_4$ solution, a mild oxidizing agent in an alkaline medium, and distilled. The nascent oxygen, liberated by $KMnO_4$, in the presence of NaOH, oxidizes the organic matter present in the soil and the ammonia released is distilled and absorbed in a known volume of standard acid and the excess of which is titrated with standard alkali using methyl red as the indicator. This method is the quickest of all the other methods for the estimation of available nitrogen in soil and has been found to work well in Indian soils.

3.6.2 Apparatus

1. Atomatic distillation unit
2. Kjeldahl flask
3. Burette, etc.

3.6.3 Reagent

1. 0.32% Potassium permanganate ($KMnO_4$) solution: Dissolve 3.2 g of potassium permanganate in distilled water, and make up the volume to 1 litre.
2. 2.5% Sodium hydroxide (NaOH): Dissolve 100 g of sodium hydroxide pellets in distilled water, and make up volume to 1 litre.
3. 0.02 *N* Sulphuric acid (H_2SO_4).
4. 2% Boric acid solution containing 20 ml of mixed indicator per litre.
5. Mixed indicator: 0.066 g methyl red plus 0.099 g bromocresol green dissolved in 100 ml of 95% alcohol.

3.6.4 Procedure

Take 5 g soil in a dry kjeldahl flask (tube) and add 20 ml of distilled water to moist the soil

↓

Add 50 ml of 0.32% $KMnO_4$ + 50 ml 2.5% NaOH + paraffin wax (approx. 1-2 g)

↓

In the distillation (receiving) end, keep a 250 ml conical flask containing 20 ml 2% boric acid and 5-6 drops of mixed indicator to collect the released ammonia

↓

Start distillation by using automatic distillation unit for 9-12 minutes (nearly 100 ml of distillate to be collected)

↓

Pinkish colour turns to light green by the absorption of ammonia in conical flask

↓

Titratien with 0.02 *N* H_2SO_4

↓

End point: Green colour changes to pinkish colour (initial colour)

Note: *The tube of distillation end must be dipped into boric acid solution.*

3.6.5 Calculation

$$\text{Mineralizable/Available N (kg/ha)} = \frac{V \times 0.02 \times 0.014 \times 2.24 \times 10^6}{W.S}$$

Where,

V = Volume of 0.02 N H_2SO_4 required for titration

W.S = Weight of soil sample taken

N.B.

Blank correction (without soil) is to be made for the final calculation

References

AOAC, Association of Official Agricultural Chemists (1930). Fertilizers general methods, Chapter 2. In: "Official and tentative methods of analysis of the AOAC, Washington, D.C.

Sahrawat, K.L. and Burfird, J.R. (1982). Modification of the alkaline permanganate method for assessing the availability of soil nitrogen in upland soils. *Soil Sci.* 133:53-57.

Stanford, G. (1978). Evaluation of ammonium release by alkaline permanganate extraction as an index of soil nitrogen availability. *Soil Sci.* 126:244-253.

Subbiah, B.V. and Asija, G.L. (1956). A rapid procedure for the estimation of available nitrogen in soils. *Current. Sci.* 25:259-260.

Tamhane, R.V. and Subbiah, B.V. (1962). Correlation and soil tests with pot and field trials in the evaluation of soil fertility. *Soil Sci. Plant Nutr.* 8:5-15.

3.7 Available Phosphorus

Available phosphorus determined by chemical extractions includes all forms of phosphorus, but predominantly which are associated with iron, aluminum, and calcium. The relative amounts of these forms of phosphorus brought into solution will depend partly upon the amount present in the soil and partly on the solubilities in the extractants used. Therefore, a suitable extractant should be selected to best reflect the phosphorus status of a particular soil. The calcium phosphate of soils is best extracted with acids, iron and aluminum phosphate rich soils with alkali and so on. In India, Bray P_1 (0.025 N HCl +0.03 N NH_4F) and Olsen (0.5M $NaHCO_3$, pH 8.5) extractants are used for the estimation available phosphorus in the State Soil Testing Laboratories for the recommendation of P fertilizers based on soil tests in acidic and neutral to alkaline soils, respectively.

3.7.1 Determination of phosphorus in solution by molybdenum blue method

In the ascorbic method of analysis, orthophosphates react, under acidic condition with ammonium molybdate $[(NH_4)_6Mo_7O_{24} \cdot 4H_2O]$ and potassium antimony tartarate $[K(SbO)C_4H_4O_6 \cdot 5H_2O]$ to form an antimony-phospho-molybdate complex. Reduction of this complex by ascorbic acid gives it an intense blue colour that is proportional to the ortho-phosphate concentration.

This reagent gives a stable blue colour (24 hours) within 10-minutes and is not affected by organic phosphorus and enabling reduced solutions to be kept overnight if necessary before measurement. The colorimetric determination of phosphorus in solutions with ascorbic acid is given below outlined by John (1970).

3.7.1.1 Reagents

Stock solution: Dissolve 20.0 g of ammonium molybdate in about 200 ml of water. Then add slowly 450 ml of 10N H_2SO_4 with stirring followed by 100 ml of 0.5% antimony potassium tartarate solution. Dilute the solution to one litre with water and store away from direct sunlight.

Mixed reagent: Add 1.5 g of L- ascorbic acid to 100 ml of stock solution. Prepare this reagent freshly, because it cannot be kept more than 24 hours.

Stock standard phosphorus solution (50 ppm P): Dissolve 0.2197 g of oven-dried AR grade potassium dihydrogen phosphate (KH_2PO_4) in distilled water, and dilute the solution to one litre. One ml of this solution contains 50 µg of P.

Working standard phosphorus solution (2 ppm P): Transfer 10 ml of 50 ppm P stock solution to 250 ml volumetric flak and dilute with water up to the mark and mix well. This standard solution contains 2 μg of P.

3.7.1.2 Preparation of standard curve

(a) Transfer 0.5, 1.0 2.5, 5.0, 7.5, 10.0, 12.5, 15.0, 20.0 and 25.0 ml of 2 ppm standard working P solution into 50 ml volumetric flasks and make the volume around 40 ml with distilled water. The concentration of P in 50 ml volumetric flasks would be 0.02, 0.04, 0.10, 0.20, 0.30, 0.40, 0.50, 0.60, 0.80 and 1.00 μg of P/ml.

(b) Add 5 ml of mixed reagent and make volume 50 ml with water.

(c) Mix well and leave between 30 and 60 minutes for colour development.

(d) Run one check with distilled water only and follow the above procedure for colour development.

(e) Measure the transmission at 882 nm by spectrophotometer/colorimeter.

(f) Prepare the standard curve by plotting concentration of P versus transmittance.

(g) Estimation of phosphorus in soil/plant extracts: Transfer an aliquot in 50 ml volumetric flasks and proceed as given for the standard curve. Determine the concentration of P from standard curve and calculate the concentration of P in soil/plant extracts.

P concentration (ppm P) = P concentration from curve X 50 X total volume of extract ÷ ml aliquot taken X weight of soil or plant samples

3.7.2 Determination of soil available phosphorus

3.7.2.1 Equipments

1. Photoelectric colorimeter/ spectrophotometer
2. Pipette (5 ml)
3. Volumetric flasks (50 ml)
4. Rotating or reciprocating shaker
5. Whatman No. 42 or No.44 (or equivalent) filter paper
6. Conical flaks (150 ml capacity)

3.7.2.2 Bray P_1

This method was developed by Bray and Kurtz in 1945 and is now widely used in all the parts of the world. The combination of HCl and NH_4F is designed to remove easily acid soluble P forms, largely calcium phosphates, and a portion of the aluminium and iron phosphates (Olsen and Sommers, 1982). The F^- ion of NH_4F dissolves Fe- and Al-phosphates by complexing with Fe^{3+} and Al^{3+} ions in acid solution with consequent release of P (Jackson, 1973).

$$3NH_4F + 3HF + AlPO_4 \rightarrow H_3PO_4 + (NH_4)_3 AlF_6$$

$$3NH_4F + 3HF + FePO_4 \rightarrow H_3PO_4 + (NH_4)_3 FeF_6$$

3.7.2.2.1 Reagent

Extracting Solution (0.03 N NH_4F in 0.025 N HCl): Dissolve 1.11 g of NH_4F in 400 ml of distilled water and transfer to one litre volumetric flask. Pipette 25 ml of standard 1 N HCl and transfer to the NH_4F solution flask and make the volume 1 litre with water.

3.7.2.2.2 Extraction of soil P and its estimation

- Take 5 g of soil sample in a 150 ml conical flask and add 50 ml of extracting solution.
- If it is necessary to obtain a colorless filtrate, add 1 teaspoon of P- free charcoal.

- Shake for 5 minutes on mechanical shaker and filter through Whatman No. 42 filter paper (the filtrate should be clear).
- Take 5 ml of the aliquots (extracts) in 50 ml volumetric flasks and one blank with extracting solution.
- Follow the procedure of P estimation in the extracts with ammonium molybdate and potassium antimony tartarate using ascorbic acid as a reducing agent.

3.7.2.3 Olsen's P

The "Olsen P" or sodium bicarbonate soil test phosphorus (P) method was developed by Olsen *et al.* (1954) to predict crop response to fertilizer P on calcareous soils. The Olsen P method is best suited for calcareous soils, particularly those with >2% calcium carbonate, but it is also effective for acidic soils (Fixen and Grove, 1990). The method is based on the use of the HCO_3^-, CO_3^{2-} and OH^- in the pH 8.5, 0.5 M $NaHCO_3$ solution to decrease the solution concentrations of soluble Ca^{2+} by precipitation as $CaCO_3$ and soluble Al^{3+} and Fe^{3+} by formation of Al and Fe oxyhydroxides, thus increasing P solubility. The increased surface negative charges and/or decreased number of sorption sites on Fe and Al oxide surfaces at high pH levels also enhance desorption of available P into solution (Sims, 2000).

3.7.2.3.1 Reagents

Olsen P Extracting Solution (0.5M $NaHCO_3$, pH 8.5): Dissolve 42.0 g sodium bicarbonate ($NaHCO_3$) in distilled water and make the volume to 950 ml. Adjust the pH of the solution to 8.5 with sodium hydroxide solution. Make the volume to 1 litre with distilled water.

Dargo-G 60 (activated charcoal): Make free from P by repeated leaching with $NaHCO_3$.

3.7.2.3.2. Procedure

- Weigh 2.5 g of soil samples into 100 ml conical flasks and add one tea spoon P-free charcoal.
- Add 50 ml of extracting solution to each flask and shake at 200 or more epm for 30 minutes at a room temperature at 24 to 27°C on an electric shaker.
- Filter extracts through Whatman No. 1 filter paper or through a similar grade of paper.
- Prepare a blank with all the reagents except the soil.
- Pipette out 5 ml of the extract in a 50 ml volumetric flask, and make volume 40 ml with water and follow the method of P estimation following the procedure described earlier.
- Take the readings of the samples as well as that of blank and calculate the concentration of P in the extracts.

3.7.2.4 Phosphorus Extraction with Iron Oxide-Impregnated Filter Paper (Pi test) (Chardon, 2000)

When a soil is shaken in $CaCl_2$ with a strip of a precipitate of amorphous iron (hydr) oxides (FeO) paper, P will first desorb from the soil, then adsorb onto the FeO-strip and again P will desorb from the soil to maintain the equilibrium. During shaking the strip is protected against erosion by soil particles via a polyethylene screen. After shaking, the strip is taken out and adhering soil particles are removed by rinsing with distilled water using an airbrush. The FeO on the paper with the P adsorbed onto it is dissolved in H_2SO_4 and P is determined in the acidic solution.

3.7.2.4.1 Equipment

1. 15-cm discs of ash-free, hard filter paper (Whatman No. 50)

2. Tweezers
3. Immersing baths
4. Polyethylene shaking bottles (100 ml)
5. Polyethylene screen (925 μm openings)
6. Shaking apparatus, end-over-end
7. Air brush

3.7.2.4.2 Reagents

Acidified $FeC1_3$ solution: Completely dissolve 100 g $FeC1_3$ in 110 ml concentrated HCl and dilute with distilled water to 1 litre.

5 % NH_4OH: Dilute 200 ml NH_4OH (25%) to 1litre with distilled water.

0.01 M $CaCl_2$: First prepare stock solution of 0.1 M: by dissolving 14.7 g $CaCl_2 \cdot 2\,H_2O$ in distilled water and dilute to 1 litre. Dilute this solution to tenfold in-distilled water to get 0.01 M $CaCl_2$ solution.

0.1 M H_2SO_4: Prepare stock solution of 2.5 M (140 ml of concentrated H_2SO_4/litre diluted with distilled water to 1 litre). Transfer 40 ml of 2.5 M H_2SO_4 to 1 litre in volumetric flask and dilute with water to the mark. The solution is 0.1 M H_2SO_4.

Distilled/deionizad water

3.7.2.4.3 Procedures

Preparation of FeO paper: Immerse the filter paper in acidified $FeC1_3$, using tweezers, for at least 5 minutes and drip dry at room temperature for 1 h. Pull the paper rapidly and uninterrupted through a bath containing 2.7 M NH_4OH to neutralize the $FeC1_3$ and produce amorphous iron (hydr) oxide (ferrihydrite, denoted as FeO). Rinse the paper with distilled water to remove adhering particles of FeO. After air drying, cut the paper into strips with a (reactive) surface of 40 cm^2 (generally 2 by 10 cm).

Shaking soil suspension with FeO strip added

1. Add 40 ml 0.01 M $CaCl_2$ to 1 g of soil in a 100 ml bottle; add one strip protected by polyethylene screen, in a fixed position, at room temperature.
2. Shake on a reciprocating shaker at a speed of 130 excursions/min, or at 4-rpm end over-end, for 16 h.
3. Take out the strip, thoroughly rinse with distilled water to remove adhering soil particles using an air brush, and remove adhering water.
4. Dissolve the FeO with adsorbed P by shaking 1 h in 40 ml 0.1 M H_2SO_4 and determine the P in the acidic extract with colorimeter/spectrophotometer.

FeO Extractable P (Pi value, ppm P) = Cp V/W

Where,

Cp = P concentration in H_2SO_4, mg/L,

V = volume of H_2SO_4, L,

W = mass of soil used, g.

References

Bray, R.H. and L. T. Kurtz, 1945. Determination of total, organic and available forms of phosphorus in soils. Soil Sci. 59:39-45.

Chardon, W.J. (2000). Phosphorus extraction with iron oxide-impregnated filter paper (Pi test). In: " *Methods for Phosphorus Analysis for Soils, Sediments, Residuals, and Water*" (ed. Pierzynski, G.M.), pp. 27-30. North Carolina State University, Raleigh, NC, USA.

Fixen, P.E. and Grove, J.H. (1990). Testing soils for phosphorus. P. 141-180. In R.L. Westerman (ed.) Soil Testing and Plant Analysis. SSSA, Madison, WI.

Jackson, M.L. (1973). *Soil Chemical Analysis* (Indian edition). Prentice-Hall of India Pvt. Ltd. New Delhi.

John, M.K. (1970). Colorimetric determination of phosphorus in soil and plant materials with ascorbic acid. *Soil Sci.* 109: 214-220.

Olsen, S.R. and Sommers, L.E. (1982). Phosphorus. *In: Methods of Soil Analysis,. 2nd ed. Agronomy Series No.9, Part 2.* (Eds. Page, A.L., Miller, R.H, and Keeney, D.R.), pp. 403-430. Soil Science Society of America, Inc., Madison, WI.

Olsen, S.R., Cole, C.V., Watanabe, F.S. and Dean, L.A. (1954). Estimation of available phosphorus in soils by extraction with sodium bicarbonate. *USDA Circular 939*. U.S. Government Printing Office, Washington D.C.

Sims J.S. (2000). Soil test phosphorus: Olsen P. In: Methods for P Analysis for Soils, Sediments, Residuals, and Waters (ed. Pierzynski, G.M.), pp. 20-21.

3.8 Available Potassium

The importance of potassium as an essential plant nutrient was first recognized by Home, 1962. Potassium is now known to be required by plant in large quantities and it has contributed the productivity of crops in intensive agriculture. In India, in past, it was widely held that with the exception of some commercial crops, there was little or no need of potassic fertilizers. The farmer's field trials in the fifties demonstrated that the application of potassic fertilizer was effective in improving yield and crop quality. The roles of potassium in plants on photosynthesis, carbohydrate and protein synthesis and water economy have already been confirmed. The higher level of potassium nutrition enhances amino acid translocation from the vegetative plant parts to grains, thereby promotes protein synthesis.

In soil, there exists three forms of K namely; unavailable, slowly available or fixed, and readily available or exchangeable potassium.

Unavailable or structural K: Soil K is contained within the crystalline structure of either primary or secondary minerals (micas, feldspars, and clay minerals). These forms of K can be made available to plants after weathering or breakdown and releasing their K as the available potassium ion (K^+). This process is too slow to supply the full K needs of field crops. The deep-rooted plants (tree and perennials) utilize the major portion of it from the weathering of potassium containing minerals.

Slowly non-exchangeable or fixed K: This form of K is trapped between the layers or plates of certain kinds of clay minerals. Non-exchangeable K is different from structural K in that it is not bonded covalently within the crystal structures of soil mineral particles. It is moderately too sparely available to plants, depending on various soil parameters (Goulding and Talibuddin, 1979; Sparks and Huang, 1985), and largely determines the soil's ability to supply potassium over the long period of time.

Readily available K: It includes K dissolved in soil solution or held on the surface of clay particles (exchangeable K). Potassium in solution tends to equilibrate with K^+ in the adsorbed fraction so that these soil K^+ fractions are closely interdependent. However, the equilibrium between solution and exchangeable K^+ is controlled to a large extend by the degree of K^+ selectivity of the adsorption sites in the exchangeable fraction. Most soil tests for available potassium measure the readily available forms but not the unavailable and slowly available forms.

3.8.1 Available Potassium

Available potassium incorporates both exchangeable and water soluble forms of the nutrient present in soil. The readily exchangeable plus water-soluble K is determined in the neutral normal ammonium acetate extract of soil. The ammonium ion provides a sharp and rapid separation from exchange complex while other cations bring about a gradual replacement of either lesser or greater amount of K which generally increase with the period of contact. The estimation of potassium in the extract is carried out with the help of flame photometer.

3.8.1.1 Apparatus

1. Flame photometer
2. pH meter
3. Electric shaker
4. Conical flask (250 ml), etc.

3.8.1.2 Reagents

1. *Neutral Normal Ammonium acetate (NH_4OAc):* Dissolve 77.09 g of NH_4OAc in 800 ml of distilled water and adjust the pH of the solution to 7.0 with dilute ammonia and acetic acid solution. Finally make the volume to 1 litre.
2. Standard solution of K

   Dissolve oven dried (120 ^oC for 1-2 hours) and cooled 1.907 g AR KCl in 1 litre volumetric flak with distilled water up to the mark. This solution contains 1000-ppm stock solution of K.

   Take 50 ml stock solution to 500 ml volumetric flask and dilute with distilled water to 500 ml final volume and mix well. This will be working stock solution of 100 ppm K.

   Prepare a series of standard solutions from the working stock solution as follows: Dilute 5, 10, 15, 20, 25, 30, 40, and 50 ml working stock solution to 100 ml in volumetric flasks with water or preferably the extractant used for the extraction of K. These solutions contain 5, 10, 15, 20, 25, 30, 40, and 50 ppm K, respectively with one volumetric flask for the check of extractant.

3.8.1.3 Extraction of K

Take 10 gm of soil in 250 ml conical flask

↓

Add to it 100 ml of N NH_4OAc (pH = 7) solution. (1:10 Soil: Suspension ratio)

↓

Shake the contents of the conical flask on an electric shaker for 30 minutes and filter

↓

Use it for analysis of K

3.8.1.4 Measurement of K with flame photometer

1. Flush the atomizer by feeding a high pressure air supply through its air-tube, and dip its capillary tube in distilled water.
2. Connect drain-tube to the drainpipe.
3. Connect the gas tube from LPG cylinder to the fuel gas inlet of the main unit.
4. Connect the air tube from the compressor to the atomizer.
5. The output regulator is adjusted to give output pressure of 0.45 kg/cm^2.
6. Switch on the fuel supply from the LPG cylinder and immediately ignite the flame with an Ignition Lighter through the ignition window.
7. Do fine adjustment of fuel flow with the help of fuel gas fine adjustment valve to get a stable flame having well defined cones.
8. Select the filter 768 nm for K.
9. Run the series of the prepared standard K solutions, and draw the calibration curve from the reading of the standard.
10. Take the emission readings of soil extract as with standard solutions.

11. Calculate the K concentration from the calibration curve drawn.

Extractable K (ppm) = ppm K from the standard curve X EV/Wt

Where,

EV = total volume of extract

Wt = weight of soil used for extraction

In the aliquot potassium can also be measured by Atomic Absorption Spectrometer/ICP.

3.8.2 Slowly, non-exchangeable or fixed K (Jackson, 1973)

3.8.2.1 Apparatus

150/250 ml conical flasks

Hot plate with temperature control system

Funnels

Filter paper

3.8.2.2 Reagents

Nitric acid, 1N and 0.1N

3.8.2.3 Procedure

1. Take 2.5 g soil into a 125 ml conical flask and add hot 25 ml 1N HNO_3 in it.
2. Boil the suspension gently for 10 minutes on a hot plate or ring stand over a gas burner after the starting of boiling.
3. Remove the flask, and filter the content into a 100 ml volumetric flask. Wash the soil left on filter paper with four 15 ml portions of 0.1N HNO_3.
4. Make the volume to 100 ml after cooling the solution and mix it thoroughly.

5. Estimate the potassium concentration in aliquot with flame photometer as given for available potassium (8.1.4).

6. The non-exchangeable K (fixed K) is calculated as the difference between the K extracted by boiling HNO_3 and that exchangeable (available) K with neural normal NH_4OAc.

References

Guilding, K.W.T. and Talibuddin, O. (1979). Potassium reserves in a sandy clay soil from the Saxmundhum experiment: Kinetics and equilibrium thermodynamics. *J. Soil Sci.* 30:291-302.

Jackson, M.L. (1973). *Soil Chemical Analysis* (Indian edition). Prentice-Hall of India Pvt. Ltd. New Delhi.

Sparks, D.L. and Huang, P.M. (1985). Physical chemistry of soil potassium. In " *Potassium in Agriculture*" (ed. Munson, R.D.). ASA/CSSA/SSSA, Madison, WI, USA.

3.9 Available (Exchangeable) Calcium and Magnesium

Calcium and magnesium are the two most abundant alkaline earth cations in soils. Both occur as i) soluble ions in soil solution, ii) ions in adsorbed state (exchangeable), iii) non-exchangeable ions in primary and secondary minerals and in organic materials. The extraction procedure for the adsorbed (exchangeable) Ca^{2+} and Mg^{2+} involves equilibrating the soil with neutral normal NH_4OAc solution for exchange with NH_4^+ ions in a short time. The Ca^{2+} and Mg^{2+} are determined by:

i) Complexometric titration method (EDTA method)

ii) Optical method (Flame photometry)

3.9.1 Complexometric Titration Method (EDTA Method)

The most widely used method for the combined determination of Ca and Mg is by complexometric titration, involving EDTA (ethylene diamine tetra-acetic acid). Analytical success depends on (a) eliminating the interfering ions, (b) achieving the required pH, and (c) employing the appropriate indicators.

When a sample containing Ca^{2+} and Mg^{2+} ions is treated with a solution of EDTA, Ca^{2+} ions are complexed as CaY^{-2}. After all the Ca^{2+} has been complexed in this fashion, the Mg^{2+} ions are complexed as MgY^{-2}. For the combined determination of Ca^{2+} and Mg^{2+}, Erichrome black T (EBT) is used as an indicator. The optimum pH for the formation of Mg-complex is 10, and this is maintained by a buffer solution. The Mg^{2+}-EBT coloured complex is less stable than the MgY^{-2} complex. Ca^{2+} is complexed before Mg^{2+}, and the indicator changes from red to blue, when the Mg^{2+} end point is reached

The classical indicator employed for Ca^{2+} titration is calcon. Calcon, at a pH 11, is purple in colour in the absence of Ca^{2+} but, in the presence of Ca^{2+} ions, it forms a pink complex, which has a dissociation constant greater than the CaY^{-2} complex. Thus, using calcon as an indicator, Ca^{2+} can be titrated with EDTA even in the presence of other alkaline earth ions

3.9.2 Apparatus

1. Electric shaker
2. Burette
3. Pipette
4. Conical flask, etc.

3.9.3 Reagents

1. *Buffer Solution:* Dissolve 67.5 g of NH_4Cl in 400 ml distilled water and add 570 ml of conc. NH_4OH /

ammonia solution and finally dilute to 1 litre with distilled water.

2. *0.01 N EDTA:* If ready made is not available, dissolve 2 g of EDTA and 1.05 g of $MgCl_2$. 6 H_2O in some volume of distilled water and dilute to a volume of 1 litre. Standardize this solution against standard Ca solution.

   (Standard Ca solution, 0.01 N = 0.5 g of pure $CaCO_3$ in 10 ml 3 N HCl (approx.) dilute to volume of exactly 1 litre.)

3. *Erichrome Black T (EBT) indicator:* Dissolve 0.2 g of EBT in 15 ml triethanolamine and add 2 g of hydroxylamine hydrochloride or 0.2 g EBT in 50 ml of methanol. (Prepare fresh every 2 weeks)

4. Masking reagents (Optional)

   a) *Potassium cyanide (2% KCN) solution:* Dissolve 0.2 g of potassium cyanide (KCN) in 100 ml of distilled water.

   b) *Potassium ferrocyanide (4% $K_4Fe\ (CN)_6$) solution:* Dissolve 4 g of potassium ferrocyanide ($K_4Fe\ (CN)_6$) in some volume of distilled water and dilute 100 ml distilled water.

   (N.B. These masking reagents are not much important, without adding these reagents we can carry out the experiment).

5. *Hydroxylamine hydrochloride solution:* 5 % w/v, aqueous. Prepare fresh for every week.

6. *Calcon solution:* Dissolve 0.2 g reagent in 50 ml of methanol; prepare fresh every two weeks.

7. *Sodium hydroxide 10 % or 16 % solution:* Dissolve 100 g or 160 g of NaOH in 1 litre distilled water.

3.9.4 Extraction of calcium and magnesium

Take 10 g of soil in 250 ml conical flask

↓

Add to it 100 ml of N NH_4OAc (pH = 7) solution. (1:10 Soil: Suspension ratio)

↓

Shake the contents of the conical flask on an electric shaker for 30 minutes and filter

↓

Extract ready for Ca and Mg analysis

3.9.5 Estimation for Ca + Mg

Take 25 ml of aliquot in a 150 ml conical flask

↓

Add 5 ml of buffer solution and 5 drops of EBT indicator

↓

Titrate with 0.01 N EDTA till the colour changes from *Red* to *Bright blue.*

↓

Carry out a blank in exactly the same manner, 25 ml of water instead of Ca solution

↓

Deduct the blank reading from the sample reading

3.9.5.1 Calculations

me of (Ca + Mg)/ 100 g of soil = $(V_2 - V_3) \times 0.01 \times V/ V_1 \times 100/ W$

Where, V = Volume of extract made

V1 = Volume of aliquot taken for analysis

V2 = Volume of EDTA used for the sample (titre-value)

V3 = Volume of EDTA used for blank

W = Weight of soil

3.9.6 Determination of Ca

Take 25 ml of aliquot in 150 ml conical flask

↓

Add 5 ml of 10% NaOH and 5 drops calcon indicator

↓

Titrate with 0.01 N EDTA till the colour changes from *Red* to *Bright blue*

↓

Carry out a blank in exactly the same manner, 25 ml of water instead of Ca solution.

↓

Deduct the blank reading from the sample reading.

3.9.6.1 Calculation

me of Ca/100 g soil = $(V_2 - V_3)$ X 0.01X V/ V_1 X 100/ W

Where,

V = Volume of extract made

V1 = Volume of aliquot taken for analysis

V2 = Volume of EDTA used for the sample (titre-value)

V3 = Volume of EDTA used for blank

W = Weight of soil

Therefore,

me of Mg/100 g soil = me of (Ca+ Mg) – me of Ca

3.10 Available Sulphur

Chemical tests for estimation of available S are of recent interest in soil testing work particularly in some reported sulphur deficient areas and for certain crops whose requirement of this nutrient is high enough, often exceeding that of phosphorus. Of several extracting procedures suggested for available sulphur determination, starting from distilled water for dissolving out the sulphate form to different chemical extractants, the use of phosphate solution has often been preferred for screening the soil for its sulphur fertility status. A number of other reagents have also been tried.

As regards the estimation of sulphur in the extract turbidimetric method is being most widely followed. The extent of turbidity due to fine suspension of barium sulphate formed by reacting sulphate ions with barium chloride in solution can be accurately measured by the spectrophotometer.

3.10.1 Extraction of available sulphur

- Weigh 20 g of air-dried soil and transfer it to a 250 ml conical flask.
- Add 100 ml of the extracting solution (0.15% $CaCl_2$) and shake it for 30 minutes.
- Filter the contents through a Whatman No. 1 filter paper.

3.10.2 Estimation of sulphur in aliquot

Take suitable aliquot and measure the sulphur concentration following the procedure of Combs *et al.* (1998) as given in water analysis (5.2.5.).

3.10.3 Calculation

Available S (kg/ha) = C X V1/W.S X V2/V3 X 2.24

Where,

C = ppm of sulphur from the standard curve

V1 = Volume of extractant

V2 = Final volume made

V3 = Volume of aliquot taken

W.S = Weight of the soil

3.11 Available Iron, Manganese, Copper and Zinc (DTPA Extractable) By Atomic Absorption Spectrophotometer (AAS)

3.11.1 Principle

A mild chelating agent, DTPA (Diethylene Triamine Penta Acetic acid), with calcium chloride is used in this method (Lindsay and Norvell, 1978). This extracting solution, which extracts the easily soluble zinc, iron, copper and manganese, is buffered at pH 7.3 by triethanolamine (TEA). The pH of 7.3 buffered with TEA is used to prevent excess dissolution of the trace metals. which is highly pH dependent. Calcium chloride is used to prevent the dissolution of calcium carbonate from calcareous soils. The dissolved elements in the extract are, then, measured by Atomic Absorption Spectrophotometer (AAS). In this method, the extracted sample is converted first into an atomic vapour by a flame and irradiated by the metal being sought. The absorption of the light by the vaporized samples is related to the concentration of the desired metal in it.

3.11.2 Equipments and Apparatus

1. Conical flasks (100 ml capacity)
2. Volumetric flasks (100 ml capacity)
3. Pipettes
4. pH meter
5. Atomic Absorption Spectrophotometer (AAS)
6. Horizontal Shaker
7. Whatman No. 42 filter paper

3.11.2.1 Reagents

DTPA extracting solution: 0.005 M DTPA, 0.01 M $CaCl_2$, 0.1M TEA $(HOCH_2CH_2N)_3N$ adjusted to pH 7.3 with dilute HCl- Dissolve 149.2 g TEA (AR), 19.67 g DTPA, and

14.7 g $CaCl_2$. $2H_2O$ in approximately 200 ml of pure double distilled water in 10 litres capacity of flask. Allow sufficient time for the DTPA to dissolve, and dilute to approximately 9 litre. Adjust the pH 7.3±0.05 with 1 N HCl while stirring, and dilute to 10 litres. This solution can be kept for several months for use.

3.11.3 Reagents

1. Standard solution of Zn, Cu, Fe and Mn

Prepare 100 ml stock solution of each element as below:

Element	Salt	Mol. Wt	Quantity (g) salt required for the preparation of solution 100 ppm
Zn	$ZnSO_4.7H_2O$	287.56	0.4398
Cu	$CuSO_4.5H_2O$	249.69	0.3929
Mn	$MnCl_2.4H_2O$	197.91	0.3602
Fe	$FeSO_4.7H_2O$	278.02	0.4977

From the stock solution prepare at least 10 working standards of each element with a range from 0 to 3.0 ppm for Zn Cu, and from 0 to 20 ppm for Fe and Mn. Working standard should be prepared in same medium as used for the extraction in soil samples. Measure the elements in the extract by AAS first running a series of standard of known element, then, analyzing the prepared samples.

3.11.4 Procedure

Take 10 g of soil

↓

Add 20 ml of extractant (DTPA)

↓

Continuous shaking for 2 hours

↓

Filter by Whatman No.42 filter paper
(if filtrate is cloudy, refilter as necessary)

↓

Estimation of Zn, Cu, Mn and Fe by AAS in the extract

3.11.4.1 Analysis by Atomic Absorption Spectrophotometer (AAS) with host computer (GBC 932+)

- Put on the air *compressor*.
- Clean the burner and fill the *liquid trap* with water.
- Select appropriate *Cathode lamp* in the instrument.
- Put on the software of the concerned instrument.
- Put on the Instrument and wait till the instruction INSTRUMENT READY.
- Go to METHOD and select element.
- Go to *Instrument, if wavelength* < *300*, then background correction is needed.
- Go to *Standards* and feed the standards accordingly (Prepare at least, 10 standards using DTPA for each element) in the software menu.
- Go to FILE and SAVE AS (a).
- Click SAMPLES and at nearly half the sample numbers go for RESCALE by right click.
- Go to FILE and SAVE AS (a).
- Go to RESULTS and click FILE, NEW, CREATE and save (a).
- Go to INSTRUMENT and double click the lamp compartment, then click PEAK EHT + D2.
- Wait for INSTRUMENT READY and adjust till maximum value using front and back knobs in the instrument.
- Then adjust the light path to the center of the circle of the given cleaning card.
- Put on the acetylene cylinder.
- Go to FLAME and OPTIMISE.

- *Put on the Air knob and then open the Fuel knob (the capillary should be outside the water).*
- Immediately press the *ignition button* and adjust the flame till the blue layer is very clear.
- Click PERFORM INSTRUMENT ZERO on the optimize box, and take the maximum standard and adjust till *Instrument Absorbance* is maximum by using *vertical* and *horizontal* knobs.
- Check for blank solution, the value should be zero.
- Go to RESULT and click the *green button* to start.
- Instruction will come on the screen and continue analysis of the samples after the calibration of standard solutions.
- For stopping click the *red button*.
- For Printing Go to REPORT - REPORT HEADER.

Note: *The above instruction is for GBC 932+, so the analysts are requested to go for the specific AAS operation manual.*

3.11.4.2 Calculations

a. Weight of soil = 10 g

b. Volume of DTPA extract made = 20 ml

c. Digital reading or reading on the galvanometer = T (say)

d. Concentration (ppm) of element as read from the standard curve against T (for sample) = C (say)

e. Concentration of element in the blank solution = Cb (say)

f. Dilution factor = 20/10 = 2 times

g. Available element in the soil (ppm) = (C-Cb) X 2
Available element in the soil (kg/ha) = (C-Cb) X 2 x 2.24

The interpretation of the soil available micronutrient cations determined by AAS is given in Table 7.

Table 7. Classification of soil available micronutrient cations in different Categories

Nutrient	Nutrient category, ppm		
	Low	Medium	High
Zinc	<0.6	0.6-1.0	>1.0
Manganese	<2.0	2.0-3.0	>3.0
Copper	<0.2	0.2-0.6	>0.6
Iron	<4.5	4.5-5.6	>5.6

Reference

Lindsay, W.L. and Norvell, W.A. (1978). Development of a DTPA soil test for zinc, iron, manganese, and copper. *Soil Sci. Soc. Am. J.* 42: 421-428.

3.12 Available Boron

There are so many methods to determine the available boron in soil to predict the response of crops. The hot water extraction procedure developed by Berger and Trough (1939) is most widely used. The hot water extracts from three major pools in soils, *viz.*, an organic, and an adsorbed inorganic and a soluble inorganic pool (Gupta, *et al.* 1985). If the B concentration of the saturation extract could be converted to the B concentration of the soil solution at field moisture condition, it could then be used as an index to assess B deficiency or toxicity of plant growth. Nitrates (>20 ppm as NO_3-N) as well as Fe and Mo, in excessive amounts interfere in the analysis, which must be eliminated by the evaporation of the aliquot, followed by the addition of $Ca(OH)_2$ solution to make it alkaline.

3.12.1 *Extraction of hot water soluble boron (Bingham, 1982)*

3.12.1.1 Apparatus

1. Low-B glassware
2. Water-cooled reflux condensers
3. Evaporating dishes

3.12.1.2 Reagents

1. *Calcium hydroxide [Ca $(OH)_2$] suspension:* Add 0.4 g of $Ca(OH)_2$ (reagent grade) to 100 ml of distilled water.
2. *Hydrochloric acid (HCl), 0.1 N:* Add 8.1 ml of conc. HCl (reagent grade) to approximately 900 ml of distilled water, mix, cool to room temperature, and adjust the solution volume to 1000 ml with distilled water.
3. *Calcium chloride ($CaCl_2$), 0.01 M:* Dissolve 1.11 g of anhydrous $CaCl_2$ in approximately 900 ml of distilled water, and adjust the solution volume to 1000 ml with distilled water.

3.12.1.3 Procedure

Weigh 20 g air-dry soil in a 250 ml flat-bottom flask

↓

Add 40 ml of 0.01 M $CaCl_2$ solution

↓

Attach a water-cooled reflux condenser to the flask

↓

Heat the flask to initiate the boiling

↓

Reflux the suspension for 5 minutes

↓

Cool the flask and filter the suspension

↓

Transfer 20 ml of aliquot to an evaporating dish and add 2 ml of Ca $(OH)_2$ suspension

Contd...

Evaporate the solution to dryness and destroy organic matter by gentle heating of evaporating dish

↓

Cool the dish and add 5 ml of 0.1 N HCl and triturate the residue in the dish with policeman to assure complete dissolution of residue

---

3.12.2 Estimation of B concentration

Take 1 ml of aliquot and proceed for B determination by Azomethine-H as described for water analysis (5.2.6.1)

3.12.3 Calculation

B (ppm) = ppm B from standard curve X A/W

Where, A = Total volume of extract (ml)

W = Weight of soil

References

Berger, K.C. and Troug, E. (1939). Boron determination in soils and plants. *Ind. Eng. Chem. Anal. Ed.* 11: 540-545.

Bingham, F.T (1982). Boron. In: *Methods of Soil Analysis; Part 2-Chemical and Microbiological Properties* (2nd Edition). Publisher: American Society of Agronomy, Inc. and Soil Science Society of America, Inc., Madison, Wisconsin, and USA. pp. 431-447.

Gupta, U.C., Jame, Y., Campbell, C.A., Leyshon, A.J. and Nicholaichuk, W. (1985). Boron toxicity and deficiency: A review. *Can. J. Soil Sci.* 65:381-409.

3.13 Available Molybdenum (Acid Ammonium Oxalate Method)

Plant available Mo is generally extracted with acid ammonium oxalate (Tamm's solution) at pH 3.3 (Grieg, 1953; Gupta and Mackay, 1966). Mo extracted with hot water can be better related to plant uptake as compared to oxalate soluble Mo. However, it is difficult to extract sufficient Mo in water to give reliable analytical results.

Molybdenum is extracted from the soil with a solution of ammonium oxalate-oxalic acid (pH 3.3) in 1:10 soil: solution ratio. The method is, especially, dependent upon the formation of thiocyanate complex of Mo. Following the removal of excess oxalate and organic matter, the determination of Mo requires saturating the aqueous molybdate solution with an organic extractant, followed by treatment with potassium thiocyanate (KCNS), in the presence of ferric iron. The red ferric thiocyanate complex, when reduced by stannous chloride, gives an orange-coloured Mo thiocyanate complex, the intensity of which is measured on a colorimeter, following its extraction into an immiscible organic solvent. The estimation of Mo in solution is followed as described by Hesse (1971).

3.13.1 Instrument and Apparatus

Photoelectric colorimeter or spectrophotometer; Separating funnel (borosilicate glass, 60 ml, permanently marked to contain 45 ml, with tips cut off close to the stop cock); glass-stopper flasks of 400 to 500 ml capacity; volumetric flasks and pipettes.

3.13.2 Reagent

- Redistilled water must be used throughout the analysis.
- Ammonium thiocyanate solution, 10% w/v, aqueous.
- *Iron (III) chloride solution:* 49 g / litre $FeCl_3 \cdot 6H_2O$
- *Sodium nitrate solution:* 42.5 g $NaNO_3$/100 ml
- *Tin (II) chloride solution:* 20 g $SnCl_2$ dissolved in 20 ml conc. HCl with heating, and the solution diluted to 200 ml. Prepare daily.

- Sodium sulphite solution, 20% w/v, aqueous.
- *Isopropyl ether:* Take isopropyl ether in a separating funnel adds one-tenth of its volume of a mixture of equal parts of tin (II) chloride solution, water and ammonium thiocyanate solution. Shake and discard the aqueous layer. Prepare daily.
- *Extraction solution:* Dissolve 24.9 g of ammonium oxalate monohydrate, $(COONH_4)_2\ H_2O$, and 12.6 g of oxalic acid dihydrate, $(COOH_2)_2 . 2H_2O$, in distilled water in a volumetric flask. The pH of this solution must be 3.3 ± 0.1, adjusted with oxalic acid.
- *Standard molybdenum solution:* Dissolve 0.150 g MoO_3 in 100 ml of 0.1 M NaOH solution. Make acid with HCl and dilute to 1 litre. The solution contains 100 µg Mo.

3.13.3 Extraction of Available Mo

- Weigh 25 g of soil in a glass-stopper flask, and add 250 ml of acid ammonium oxalate and shake for 16 hours on a mechanical shaker.
- Filter the suspension through filter paper fitted on funnel, washed initially with 6.5N HCl, and discard the first 10 to 15 ml of filtrate.

3.13.4 Determination of Mo in Solution

- Take 50 ml to 100 ml of extracted test solution in a 150 ml separating funnel.
- Add 1 ml of iron (III) chloride solution, 1 ml of sodium nitrate solution, 5 ml of ammonium thiocyanate, 2 ml of tin (II) chloride solution and 1 ml of sodium sulphite solution in that order.
- Shake the solution well after each addition.
- Add 10 ml of isopropyl ether and shake for 1 minute.

- Dry the organic phase by running it through a cotton-wool plug and read its optical density at 475 nm.
- A calibration curve is prepared to cover the range 0-1 $\mu g\ ml^{-1}$ Mo.

Note: *For total molybdenum determinations the addition of iron (III) chloride can usually be omitted as sufficient iron will be present.*

3.13.5 Observations

a. Weight of soil taken = 25 g

b. Volume of the ammonium oxalate extract = 250 ml

c. Volume of extract taken = V ml

d. Volume of the aliquot made up to following various treatments = V_1 ml

e. Transmittance (%T) as read from the colorimeter = T (say)

f. Concentration (ppm) of Mo against T = C (say)

3.13.6 Calculations

g. First dilution = (250/25) = 10 times

h. Second dilution = (V_1/V)

Total dilution = $10 \times (V_1/V)$

j. Now, Available Mo content of soil (ppm) = $C \times 10 \times (V_1/V)$

3.13.7 Interpretation

Plant available Mo (ppm)	Comments
< 0.04	Deficient
0.04 – 0.2	Normal
> 0.2	Toxic

References

Grieg, J.L. (1953). Determination of available molybdenum of soils. N.Z. *J. Sci. Tech. Sect.* A-34: 405-414.

Gupta, U.C. and Mackay, D.C. (1966). Procedure for the determination of exchangeable copper and molybdenum in podzol soils. *Soil Sci.* 101:93-97.

Hesse, P.R. (1971). *A Text Book of Soil Chemical Analysis*. John Murray (publishers) Ltd. 50. Albemarle Street, London, WIX4BD. pp. 416-417.

3.14 Available Nutrients and Heavy Metals By AB-DTPA Soil Test and ICPS

The ammonium bicarbonate-diethylene triamine penta acetic acid (DTPA) (AB-DTPA) soil test was developed by Soltanpour and Schwab (1977) to extract simultaneously labile NO_3-N, P, K, Zn, Fe, Mn, and Cu from neutral and calcareous soils. Thereafter, it was used to extract the potentially toxic elements from soils and mine spoils and soil treated with sewage sludge. The bio-availability determined with AB-DTPA also includes of Pb, Cd, Ni, Se, As, B, Mo and S (Soltanpour, 1991). The estimation of elements in AB-DTPA extract is more efficiently estimated by using inductivity couple plasma spectrophotometer (ICPS). This instrument is capable of simultaneous or sequential determination of the above elements accurately and rapidly in the AB-DTPA extract. Boon and Soltanpour (1983) and Boon (1985) found that AB-DTPA extractable Pb and Cd in soil contaminated with silver mine spoil material highly correlated with their respective plant contaminations in spinach and lettuce leaves. They also observed that the AB-DTPA test is good indicator of soil contamination with Pb, Cd, Zn, and Mo. Barbarick and Workman (1987) showed a high correlation between AB-DTPA extractable Ni and Cd level, and concentration of these metals in leaves of Swiss chard grown in sludge.

Selenium in contaminated soil poses a hazard for wildlife and humans. The Se extracted by AB-DTPA (Soltanpour *et al.*, 1982) showed a high degree of correlation between Se uptakes by plants.

3.14.1 Reagents

Preparation of NH_4HCO_3-DTPA

1. Dissolve 1.97g DTPA in 800ml of pure water.
2. Add about 2 ml of 1:1 NH_4OH to facilitate dissolution.
3. After DTPA is in solution, add 79.06 g of NH_4HCO_3, and stir gently until dissolved.
4. Adjust pH to 7.6 using either NH_4OH or HCl.
5. Bring to a final volume of 1 liter.
6. The extracting solution is unstable with regard to pH and should be made immediately before using or kept under mineral oil.

3.14.2 Procedure

1. Place 10 g of soil in 250 ml Erlenmeyer (Conical) flask.
2. Add 20 ml of extracting solution and shake on a reciprocating shaker at 180 cycle per minute for 15 minutes. The flasks are kept open.
3. Filter through a medium- speed filter paper.
4. Place 0.25 ml of concentrated HNO_3 in a 10 ml beaker.
5. Carefully add 2.5 ml of AB-DTPA extract or standard solution (Table 8) to the beaker and mix on a rotary shaker for 15 min. to eliminate carbonate species.
6. Analyze on ICPS.
7. Ultra trace levels of As, Se, and Hg are determined by a hydride generator.
8. The interpretation of AB-DTPA-extractable plant nutrients and some other nutrients in soil are given in table 9.

Table 8. ICPS standard solutions (Soltanpour, 1991)

Elements	Soil (µg/g)	Concentration in AB-DTPA high standard (µg/ml)	ml stock solution/500 ml of AB-DTPA high standard
P	20	10	5
K	200	100	25
Zn	10	5	2.5
Fe	10	5	2.5
Mn	10	5	2.5
Cu	10	5	2.5
Pb	2	1	0.5
Cd	2	1	0.5
Ni	2	1	0.5
Mo	2	1	0.5
B	2	1	0.5
As	2	1	0.5
Se	2	1	0.5
Ca	200	100	25
Mg	200	100	25
Na	200	100	25
Cr	2	1	0.5
Sr	2	1	0.5
Ba	2	1	0.5
Al	2	1	0.5
Ti	2	1	0.5
Hg	2	1	0.5

Table 9. Interpretation of nutrients extracted by AB-DTPA extractant (Soltanpour, 1991)

Element	Low	Medium	High	Very high	Potential toxic range
	mg/kg (ppm)				
NO_3-N	0-10	11-20	21-30	>30	
P (light soils)	0-5	6-10	11-15	>15	
P (loamy soils)	0-2.5	3-5	5.5-8	>8	
P (clayey soils)	0-1.5	2-3	3.5-5	>5	
K	0-60	61-120	121-180	>80	
Zn	0-0.9	1.0-1.5	>1.5	-	70
Fe	0-3.0	3.1-5.0	>5.0	-	
Mn	0-0.5	0.6-1.0	>1.0	-	
Cu	0-0.2	0.3-0.5	>0.5	-	
B					5
Pb					20-100
Cd					>0.1
As					>2.5
Mo					>1.7

References

Barbarick, K.A. and Workman, S.M. (1987). Ammonium bicarbonate-DTPA and DTPA extractions of sludge-amended soils. *J. Environ. Quality* 16:125-130.

Boon, D.Y. (1985). Lead, cadmium and zinc: Evaluation on concentration using the NH_4CO_3-DTPA soil test. Colorado State Univ., Fort Collins M.Sc. (Thesis quoted by Soltanpour, 1991).

Soltanpour, P.N. and Schwab, A.P. (1979). A new soil test for simultaneous extraction of macro-and micro-nutrients in alkaline soils. *Comm. Soil Sci. Plant Anal.* 8:195-207.

Soltanpour, P.N., Olsen, S.R. and Goos, R.J. (1982). Effect of nitrogen fertilization of dryland wheat on grain selenium concentration. *Soil Sci. Soc. Am. J.* 46:430-433.

Soltanpour, P.N. (1991). Determination of nutrient availability and elemental toxicity by AB-DTPA soil test and ICPS. *Adv. Soil Sci.* 16:165-190.

4

Plant Analysis

4.1 Sampling and Preparation for Analysis

The determination of nutritional need of crops is an important aspect of nutrient management of any plant analysis programme. The need of fruit plants for mineral nutrients differ from those of annual crop species in a number of ways, many of which are related to the perennial nature of tree crops. In perennial tree crops, there is a need to supply nutrients both for ultimate fruit production, the fruits and the vegetative organs, which persist from year to year. The need for leaf analysis in perennial horticultural crops and forest trees has proved its superiority over soil diagnostic methods. This is essentially because of their deeper root system and the fact that nutrients supplied in one year may have its effect on both the nutrition and crop production in following years. However, it is largely being used not as an alternative method to soil testing, but as a supplement to soil testing. Results of foliar analysis are usually interpreted using the critical nutrient level (CNL) approach. Accurate interpretation of foliar analysis using the CNL approach is possible only when sampling is restricted to the same growth stage at which standard reference values have been established (Beaufils, 1971, 1973). The general purposes of plant analyses are:

1. To diagnose or confirm diagnosis of visible symptoms.
2. To identify hidden hunger.
3. To locate areas of incipient deficiencies.
4. To indicate whether applied nutrient has entered the plant.
5. To indicate interactions or antagonisms among nutrients.
6. To aid the understanding of internal plant functioning.
7. To suggest additional test to identify the trouble.

The element nutrients of the plant tissues can be estimated in the extract of plant obtained from the fresh plant tissues (i.e. tissue analysis), and in whole or parts of the dried plant material. The tissue test is qualitative and only valid for the quick guidance of the growing crops. Total nutrient analysis of the plant is quantitative in nature and is more reliable and uses to categorize the ranges of deficiency, low, sufficiency, and excess of the nutrient concentrations in the crops grown. The plant tests are divided into tissue tests, petiole analysis or total analysis.

Plant tissue analysis can provide on-the spot field tests. Kits vary from a few chemicals and test papers to elaborate supplies of chemicals, vials and spot plates, which are better to use indoor. Nutrient levels determined from these tests are accurate only in terms of high, medium or low values. The most tissue tests are available for N, P and K in such crops as maize, potatoes, small grains, sorghum, tomatoes and some pasture grasses. The leguminous crops are tested for P and K only. The critical concentration can be directly determined only from the relationship of the nutrient concentration to the yield or quality. The tissue should be chosen which best shows their relationship through regression.

The tissue, which is best, suited for determination of deficiencies is not always best for toxicity of nutrient elements. For a number of nutrients, the concentration in the roots may be abnormally high and limit growth, while the concentration in the tops may be in normal range (e.g. Cu, Na, and Al). The practical problem involves in sampling and cleaning of roots limit the analysis.

Leaf analysis can be a very useful tool for plant nutritional diagnosis, since adequate procedures are available for data analysis. Because of the dynamic nature of the leaf tissue composition, strongly influenced by leaf age, maturation stage, and the interactions involving nutrients absorption and translocation, the tissue diagnosis may be a practice of difficult understanding and utilization (Walworth and Sumner, 1987). Several methods for nutritional diagnosis using leaf tissue analysis have been proposed and used, including the critical value (CV), the sufficiency range approach (SRA), and the diagnosis and recommendation integrated system (DRIS). Considering the DRIS uses, the nutritional balancing concept (relationship among nutrients), it is proposed that this method might be more precise than the others in the detection of nutritional deficiencies or/and excesses.

In general DRIS has some advantage over other diagnosis methods; presents continuous scale and easy interpretation; allows nutrient classification from the most deficient up to most excessive; can detect cases of yield limiting due to nutrient imbalance, even when none of the nutrient is below the critical level; and finally, allows to diagnose the total plant nutritional balance, through an unbalance index (Baldock and Schulte, 1996). The additional advantage of DRIS, it is less sensitive to tissue aging in comparison to others (Walworth and Sumner, 1987). The criteria for critical levels or sufficiency ranges generally depend on norms for diagnosis derived from an specific plant tissue part and age, and classifies the plants based solely in the leaf nutrient concentration (leaf nutrient content/leaf dry matter). Thus, the plant growth stage for leaf sampling is an essential factor for the application of both methods, and therefore, the diagnoses based on these criteria are usually applied in leaf samples obtained from a well-defined growth stage

Petiole analysis tests are frequently used for N only or N and P levels. It gives the accuracy values between tissue testing and total plant analysis. The petiole analysis have been successfully used to monitor the N levels in crops, such as tomatoes, potatoes, grapes, cotton, sugar beets and lettuce, because the quality and yield of these crops are frequently affected by N content.

Total analysis gives the accurate status of nutrient across a variety of crops. The analyzed nutrient concentrations are reported in ppm (part per million) for micronutrients and in for primary and secondary in percentage. The total plant analysis can also be used if toxic levels of certain non-nutrient elements present in plant. Visual and analytical diagnosis during plant growth and development provide two complementary methods for identifying nutritional disorders of plants, which permit them to be distinguished from phytopathological damage too.

4.1.1 Sampling

Among the several tissues to be considered for nutritional diagnosis purposes, leaves constitute the main plant sampling material (Chapman and Brown, 1950). The improvement in tissue analysis techniques enabled the comparison of results from different soil type fertilization experiments. Therefore, for plant analysis to be meaningful, collection of particular plant part (tissues) at the right stage of growth as per technical specifications is very important. It would be wrong and wasteful to just pluck any leaf or branch from a growing plant at any time and send them to a laboratory for analysis. No amount of analytical skill or sophisticated instrumentation can take the place of correct sampling methodology. The deviation in the sampling procedure may be necessary under specific circumstances such as occurrence of deficiency/toxic symptoms or for nutrient uptake. The index tissue sampling procedure is given in table 10 is designed for nutrient diagnosis, monitoring and efficient nutrient management for economic optimum yield and for excellent quality.

4.1.2 Sample Preparation

After sample collection, the fresh tissues should be decontaminated from dust and other foreign material by adopting the following procedure. Plastic containers are taken in which the following solutions are added.

1. Wash in 0.2% liquid detergent to remove waxy coating on leaf surface + soil particle.
2. Dip in 0.1N HCl (8 ml conc. HCl/litre) in plastic container.
3. Rinse with tap water thoroughly.
4. Finally wash with distilled water.

The extra moisture is wiped out; the sample is placed in new paper bags and dried in an oven at 60-70°C. In the laboratory the samples are homogenized using a sample mill, which should not give any metallic contamination.

The plant samples (grain and straw/stalk) of mature crops should be taken in such a way that can represent the average content of nutrients in the samples. Dry the collected samples in the hot air oven at 60-70 °C for complete drying. Grind the dried samples in a grinding mill and sieved through 2 mm sieve and keep in polythene bags or bottles for mineral analysis.

Table 10. Plant tissue sampling guidelines for different crops

S/ No.	Crop	Index tissue	Growth stage	Sample size
Food crops				
1	Rice	3rd leaf from apex	Tillering	50
2	Maize	The leaves below and opposite to ear	Tasseling to silking	15
3	Pulses	Recently matured leaf	Bloom initiation	50
4	Potato	Most recently fully developed leaf	Half grown	15
Oilseed crops				
5	Groundnut	Recently matured leaflets	Maximum tillering	25
6	Soybean	3rd leaf from top	2 months after planting	25
7	Mustard	Recently matured leaf	Bloom initiation	20
8	Sunflower	Youngest mature leaf	Initiation of flowering	20
Tropical fruit crops				
9	Banana	Petiole of 3rd open leaf from apex	Bud differentiation 4 months after planting	15
10	Grapes	5th petiole from base	Bud differentiation stage for yield forecast	200
11	Citrus	3-5 month old leaf from March flush	Mature leaves from the non-bearing terminals	30
12	Guava	3rd pair of recently matured leaves	Bloom stage (Aug.-Dec)	25
13	Mango	Leaves + petiole	4-7 months old leaves from middle of shoot	15
14	Papaya	6th petiole from apex	6 months after planting bloom	25
15	Passion fruit	Matured leaf opposite to last open flower	Bloom	25
16	Pineapple	Middle one third portion of white basal portion of 4th leaf from apex	4-6months	100
Vegetables				
17	Bean	Upper most recent fully developed trifoliate leaves		10
18	Cabbage	Wrapped leaf	2-3months old	15
19	Carrot	Most recent fully developed leaf	Mid grown	15
20	Cauliflower	Most recently fully developed leaf	At heading	12
21	Brinjal	Leaf blade with midrib minus petioles fro most recently matured leaves		12
22	Garlic	Most recent fully developed leaf	Pre-bulb	12
23	Onion	Top no white portions	1/3 to [illegible]	12
24	Peas	Leaflets from most recent fully developed leaflets	First bloom	50
25	Tomato	Leaves adjacent to inflorescence	Midbloom	15

References

Baldock, J.O. and Schulte, E.E. (1996). Plant analysis with standardized scores combines DRIS and sufficiency range approaches for corn. *Agron. J.* 88:448-456.

Beaufils, E.R. (1971). Physiological diagnosis-a guide for improving maize production based on principles developed for rubber trees. *Fert. Soc. South Africa J.* 1:1-30.

Beaufils, E.R. (1973). Diagnosis and recommendation integrated system (DRIS). A general scheme of experimentation based on principles developed research in plant nutrition. *Univ. of Natal Soil Sci. Bull. 1*, Pietermaritzburg, South Africa.

Chapman, H.D. and Brown, S.M. (1950). Analysis of orange leaves for diagnosing nutrient status with reference to potassium. *Hilgardia* 19:501-540.

Walworth, J.L. and Sumner, M.E. (1987). The Diagnosis and Recommendation Integrated System (DRIS). *Adv. Soil Sci.* 6:149-188.

4.2 Extraction of Total Plant Nutrients and Their Estimation

The total concentration of plant nutrients can be determined either by wet digestion procedure or by dry ashing procedure. Although dry ashing is simpler, easier, non-hazardous and economical option, yet wet digestion is most common. Wet digestion with $HClO_4$ avoids loss of K through volatilization and gives a clear solution of all constituents (Jackson 1973) except Si, which is quantitatively dehydrated and precipitated and is removed by filtration or centrifugation. However, at higher temperature (> 200°C), H_3PO_4 may get lost during digestion. The common practice of slow heating of the digestion mixture containing HNO_3, H_2SO_4, and $HClO_4$ from room temperature to 203 °C , the constant boiling point of the mixture result the complete oxidation of $HClO_4$.

4.2.1 Apparatus

Block digestion system (Pelican equipment, Chennai, India) with temperature controller to hold the tube capacity of 250 ml, fume exhaust, suction source and safety feature.

4.2.2 Reagent

Digestion extractant: Ternary mixture of concentrated HNO_3, $HClO_4$ (60-70%) and H_2SO_4 in the ration 10:4:1 on volume basis in any quantity and allow to cool before use.

4.2.3 Procedure

Weigh 1 g oven dried (60°C) groind plant samples passed through 2 mm sieve in the dried 250 ml digestion tubes and add 10-20 ml ternary acid mixture. Swirl the tubes to moisten entire plant sample. Blank (duplicate) tubes may be kept for the reagents, add same quantity of digestion mixtures as employed for the plant samples. Thereafter, digestion is carried out at 180°C-200°C until the clear solution remains after the fumes of acid liquid almost volatized. When the residues of tubes are clear and white, it is ready for the analysis of plant mineral element analysis. Cool the digestion tubes and add distilled water from the wash bottle approximately 10-15 ml and filter the liquid using high quality filter paper in funnels fitted on 100 ml volumetric flasks. Repeat the washing of the content of tubes with distilled water and filter it until all the contents are transferred. Make the volume up to 100 ml with distilled water. These aliquots are ready for the determination of all the elements excepting N, S and B. The separate system must be employed for the estimation of sulphur in plant samples with wet digestion method without H_2SO_4 similarly as stated above.

4.2.4 Extraction of boron

The estimation of boron in plant samples is done by dry-ashing in muffle furnace. However, there is a chance of boron contamination from muffle furnace. Yoshida and Yoshida (1965) extracted the boron content of plant samples with 0.5N HCl. Paul and Smith (1966) recommended the 25 ml HCl for extraction in place 10 ml suggested by the previous authors. This method of acid extraction is convenient as compared to dry-ashing for a large number of samples.

4.2.4.1 Procedure

1. Take 0.15 to 0.25 g dried; ground, and sieved plant material in 150 ml PVC conical flasks free from B contamination.
2. Add 25 ml 0.5 N HCl in to conical flasks containing plant samples.
3. Shake two hours on mechanical shaker.
4. Filter the aliquot in 100 ml volumetric flask and wash the content of flask 3-4 times with 10 ml water and transfer the same on filter paper. Finally make the volume to 100 ml.

4.2.5 Total Nitrogen

A known quantity of powdered sample is digested with concentrated H_2SO_4 in the presence of digestion mixture by heating in a fume chamber. Under these conditions, $(NH_4)_2\ SO_4$ is formed from protein, nitrogen and non-nitrogenous compounds. The mixture is distilled and NH_3 released is passed into boric acid, which is then back titrated with standard acid.

$$\text{Protein} + \text{Non-protein compounds} \xrightarrow[\text{Conc.}H_2SO_4 + \text{Digestion mixture}]{\text{Digestion}} (NH_4)_2\ SO_4$$

$$(NH_4)_2SO_4 + 2NaOH \rightarrow 2NH_3 + Na_2SO_4 + 2H_2O$$

$$NH_3 + H_3BO_3 \rightarrow NH_4 + H_2\ BO_3^-$$

$$2NH_4 + H_2BO_3^- + H_2SO_4 \rightarrow (NH_4)_2SO_4 + 2\ H_3BO_3$$

Using the above relationship, the concentration of N in the plant sample is calculated

1ml of 1N H_2SO_4 = 0.014 g N

4.2.5.1 Apparatus

1. Kjeldahl digestion assembly
2. Kjeldahl flasks/tubes
3. Distillation assembly
4. Conical flask or beaker
5. Pipette, etc.

4.2.5.2 Reagents

1. *Concentrated* H_2SO_4
2. *40% NaOH solution:* Dissolve 400 g of NaOH in about 800 ml distilled water and make up the volume to 1000 ml.
3. *Mixed indicator:* Mix 0.1g of bromocresol green with 0.07 g of methyl red and dissolve this mixture in 100 ml of ethanol.
4. *2% boric acid:* Dissolve 20 g of boric acid in some volume of distilled water and make up the volume to 1000 ml.
5. *Standard H_2SO_4, 0.1 N:* Dilute 2.8 ml of conc. H_2SO_4 to 1000 ml with distilled water and then standardise it using standard Na_2CO_3 solution.
6. *Digestion mixture:* Mix 100 parts of K_2SO_4, 20 parts of $CuSO_4.5H_2O$ and 1 part of selenium powder and grained it to get fine powder.
7. *Standard Na_2CO_3, 0.1 N solution:* Dissolve 5.3 g of dry Na_2CO_3 in some volume of distilled water (500 ml) and then make up the volume to 1 litre.

4.2.5.3 Procedure

- Weigh 1.0 g plant sample into a digestion tube of 250 ml capacity.
- Add 5 g potassium sulphate catalyst mixture followed by 30 ml concentrated H_2SO_4.
- Shake the solution and allow standing until all the sediment has settled.
- Gently heat the tube on the digestion block cover until frothing has ceased.
- Increase the temperature until the digest gets clear (light bluish green colour).
- Gently boil the clear digest for 2-5 hours.
- After cooling, carefully add around 20 ml water, shake the solution and allow to stand for the sediment to get settle.
- Take 20-30 ml boric acid and indicator mixture in 250 ml conical flask and dip the receiving end of distillation in the solution.
- Fit the digested tube to the distillation set (Kjel-Plus)
- Add 10 ml NaOH (10 N) in increments until the colour of digest change to brown.
- Run the distillation set for 5-10 minutes by setting timer.
- Take out the 250 ml conical of boric acid trapped the released ammonia.
- Titrate the distillate with 0.10 N H_2SO_4 or HCl until the colour changes green to pink.
- Run one blank without plant sample for final calculation.
- 1 ml 0.10 N H_2SO_4 = 1.4 mg N

Note: *To include the nitrate and nitrite add 3 ml sulphuric-salicylic-acid mixture (3.4 g salicylic acid dissolved in 100 ml of the sulphuric acid) and add 0. 5 g sodium thiosulphate ($Na_2S_2O_3$) just before heating the mixture).*

4.2.6 Determination of Phosphorus by Vanadomolybdophosphoric Yellow Colour Method in Nitric Acid (Jackson, 1973)

This method has the advantages of colour stability, freedom from the reduction step, greater freedom from interfering ions, and contamination from glassware and lower dilution required for a convenient size of soil and plant sample analyses. It is one-tenth sensitive than molybdenum blue method. The development of yellow colour has been attributed to substitution of oxyvanadium and oxymolybdenum radicals for the O of PO_4 to give a heteropoly compound of the colour.

4.2.6.1 Apparatus

50 ml volumetric flasks, pipettes, 1 litre volumetric flask, colorimeter/spectrophotometer, etc.

4.2.6.2 Reagents

Vanadomolybdate reagent: Dissolve 25 g ammonium molybdate in 400 ml water. Then dissolve separately 1.25 g ammonium metavanadate in 300 ml of boiling water. After cooling of the ammonium metavanadate solution, add 250 ml of concentrated HNO_3 in it and again cool the solution to room temperature. Transfer the mixture in 1 litre volumetric flask. Finally add the ammonium molybdate solution into this and dilute the mixture to 1 litre.

50-ppm standard P stock solution: Prepare as given earlier for available P estimation (3.7).

4.2.6.3 Development of colour in standard curve preparation and test solution

Pipette 1, 2.5, 5.0, 7.5, 12.5. 15.0, and 20.0 ml of 50-ppm solution into 50 ml volumetric flasks. Add 10 ml of the vanadomolybdate reagent. Dilute the solution with distilled water up to 50 ml mark and mixe thoroughly. A blank must be prepared to exclude the impurity of chemicals in respect of phosphorus contained in. The colour develops rapidly and take reading after 10 minutes with the adjustment of colorimeter/spectrometer at 100% transmission or optical density at zero with blank at 400 to 490 nm according to sensitivity needed for different light maxima:

Range, ppm P	Light maximum, nm
0.75-5.5	400
2.0-15.0	440
4.0-17.0	470
7-20.0	490

A suitable selection of light maxima depends on the taken test solution according to the transmission percentage found. Prepare the standard curve by plotting concentration of P vs % transmission/optical density. Take the suitable aliquot of plant-digested samples, similarly estimate the concentration of P.

P concentration (ppm P) = P concentration from curve X 50 X total volume of extract ÷ ml aliquot taken X weight of plant samples for digestion

4.2.7 Estimation of K, micronutrient cations, boron, calcium, magnesium, sulphur and potassium

In the plant extracts these elements can be estimated following the methods described in soil and water analysis for the respective element (3 and 5). The interfering ions

Fe, Mn, and Cu in the extract for the estimation of Ca and Mg by EDTA can be eliminated with diethyl-dithiocarbamate, which forms brown complexes with these metals that are removed into iso-amyl alcohol leaving calcium and magnesium in the aqueous phase.

References

Jackson, M.L. (1973). *Soil Chemical Analysis* (Indian edition). Prentice Hall of India Pvt. Ltd. New Delhi.

Paul, J.L. and Smith, L.V. (1966). Note: Boron analysis in plant material. *Soil Sci.* 102:353.

Yoshida, Y. and Yoshida, S. (1965). Rapid determination of boron in plant tissue. *J. Soil Sci. Manure* 36:45-48.

—•—

5

Water Analysis

5.1 Sampling

There are different sources of water such as rain, streams, canals, lakes, tanks, ponds, large reservoirs, shallow and deep wells, shallow and deep tube wells, as well as sewage effluents and sea water. The composition of the different sources of water vary with the season, rate of flow, the extent of effluents from the industrial units and sewage effluents from the municipal area during rainy season, the river tank and pond water becomes salty and muddy due to soil erosion. The depth of ground water also varies with the seasons. The sources of water available to the farmer for irrigation of crops are generally few such as canal, shallow and deep wells, pumping sets, tube wells and run-off water collected in excavations along roads, natural drains and adjoining areas under impeded drainage. The composition of water also varies with the season. Whatever may be sources of water, the water sample is taken along with the soil sample to soil testing laboratory for its analysis by way of assessing the suitability of water for irrigation.

For drawing maximum benefit the samples taken be representative of the water in bulk and it should be collected in a clean vessel. One to three litres capacity plastic bottles duly washed with a detergent are satisfactory. However, before filling with water samples the bottles must be rinsed with water samples three times. The information such as farmers name, owner of pumping sets, date and month of sampling depth of ground water, crops and soils being irrigated, any specific agricultural or human problems encountered to be briefly recorded along with the name of village, block, tehsil and district. The water sample should be properly labeled and submitted to soil testing laboratory for its analysis.

5.1.1 Sampling Tools and Accessories

1. Clean plastic or glass bottles preferably of 500ml capacity
2. Cotton or nylon thread
3. Water table recorder
4. Glass marking pencils and
5. Labels and tags

5.1.2 Sampling of Water

Water sample is collected in the bottle after rinsing it with the same water 3-4 times. Sample from tube well should be collected directly from the outlet point. When sampling from open wells and peizometers, the sample bottle may be tied to a good quality thread and then lowered into the well till it is filled with water. Care should be taken to collect the sample only after continuous discharge of the sources for 10-20 minutes. From shallow ponds, lakes, reservoirs, canals and rivers, water sample should be collected form the centre of the width after avoiding the floating debris on the top. About half a litre of the sample is sufficient. After proper marking and labeling immediately the sample should be sent to the laboratory for analysis in order to avoid any change/deterioration in its quality due to chemical or microbial activity.

5.1.3 *Precautions to be Taken*

1. Water sample should not be collected in dirty, used and contaminated bottles. The container should be transparent and preferably new.
2. Time interval between collection and analysis of the sample should be short in order to get reliable results.
3. If delay is inevitable, water sample must be protected against bacterial growth either by adding 2-3 drops of pure toluene or by keeping it under refrigeration for the time required.

5.1.4 *The Parameters Generally used to Assess Water Quality*

Physical : Temperature, colour, odour, turbidity, foam and froth, and suspended solids.

Chemical: BOD, COD, DO, pH, hardness, total salt, cations/anions, detergent and pesticides.

Biological: Bacteria, coliforms, algae, and virus.

In a laboratory the water samples are generally analyzed for pH, EC, concentration of cations and anions from the agricultural point of view.

5.2 Determination of Water Quality

5.2.1 *Total Salts*

The concentration of total salts in water samples can be estimated most conveniently from their electrical conductance. Conductivity is reported as micromhos/ cm (mmhos/cm) or EC X 10^6. Now the results of conductivity is expressed as desisiemens per meter (dS/m = mmhos/cm)) from the electrical conductivity measurement, table can be used to derive mg/ L or ppm salt. It is measured with the help of an electrical conductivity meter.

5.2.1.1 Procedure

Needed apparatus consists of an AC " salt bridge" or "electrical resistance bridge", conductance cell with platinum-blackened electrodes and vacuum tube. Take the water sample in clean beaker (100 ml). Rinse the conductivity cell with distilled water and then with the sample. Temperature and cell constant corrections are adjusted on the conductivity meter if provided. Connect conductivity cell to meter and dip in the sample. Pass the current and adjust the current by rotating the dial in such a way that the width of shadow in cathode ray is maximum. Then read the conductivity value in dSm. The salt bridge with its conventional cell reads directly in specific conductance, no specific constant is needed. For too, concentrated solutions an aliquot should be diluted ten times and the obtained reading multiplied by ten.

ppm of salts = 640 X dS/m

% Salts in solution = 0.064 X dS/m

Another method of determining the total salinity is by chemical analysis. The basis of the method is to evaporate off the water and determine the total salts gravimetrically. With saline water, it is not a problem to get the clean water, however, it is difficult to get the clean water. The summation of individual ions provides another method of determining total soluble salts.

5.2.2 pH

The pH of the water determines whether the water is acidic or basic (alkaline). The pH of natural waters ranges from 6.5 to 8.5. Industrial and community wastes, acid rain, bedrock type, and the biological processes of photosynthesis, respiration and decomposition all influence pH levels. For saline irrigation water, determination of pH is not very important since presence of neutral salts keeps it around 7.0. However the pH of bicarbonate (HCO_3^-)

water is usually more than 7.5 and its determination may reflect the degree of sodicity in the sample.

5.2.2.1 Procedure

The pH is determined by taking about 50 ml water in a 100 ml clean beaker and immersing the glass and calomel electrodes or combined electrode of the pH meter in it. The pH is indicated on the dial. pH meter is first adjusted to the known pH of buffer solutions having pH of 4.0 and 9.2.

5.2.3 Carbonates

The usual method of determination in water is by titration with dilute acid, firstly to bicarbonate (HCO_3^-) stage using phenolphthalein as indicator,

$$CO_3^{-2} + H^+ \quad = \quad HCO_3^-$$

and then, using the same aliquot, to the carbonate stage using methyl orange or other suitable indicators,

$$HCO_3^{-2} + H^+ \quad = \quad H_2CO_3 \leftrightarrow CO_2 \; + \; H_2O$$

5.2.3.1 Reagents

1. *Standard sulphuric acid (0.10 N):* Take 2.8 ml of concentrated H_2SO_4 (36 N) with automatic pipette and dilute to one litre with distilled water. Standardize with 0.10 N standard solution of Na_2CO_3.
2. *Phenolphthalein indicator (0.25%):* Dissolve 0.25 g of pure phenolphthalein powder in 100 ml of 60% alcohol.
3. *Methyl orange indicator (0.5%):* Dissolve 0.5 g dry methyl orange powder in 100 ml of 95% alcohol.

5.2.3.2 Procedure (Hesse, 1971)

Pipette a suitable quantity of water (50 ml) into a white porcelain basin and add a few drops of phenolphthalein indicator. Titrate the solution with 0.10 N standard sulphuric acid until the pink colour disappears and note the volume of acid used. Add 3-4 drops of methyl orange solution and continue the titration until the indicator turn red.

$$HCO_3^- \text{ me/L} = \frac{\text{(A-2B) X normality of acid X1000}}{\text{Volume of water taken (cm}^3\text{)}}$$

$$CO_3^{-2} \text{ me/L} = \frac{\text{2B X normality of acid X1000}}{\text{Volume of water taken (cm}^3\text{)}}$$

Where,

A= volume of acid used after the methyl orange solution addition.

B = volume of acid used in titration after the phenolphthalein solution addition.

The burette reading in ml can be multiplied by 2 gives the CO_3^{-2} me/L or equivalent to ppm of solution. If CO_3^{-2} is found absent, the burette reading in ml is numerically equal to HCO_3^- me/L or ppm of solution.

5.2.4 Ca and Mg

In water hardness is the measure of the total concentration of calcium and magnesium ions and to a less extent, the salt of other minerals. Calcium and magnesium enter the water via the action of carbonic acid. As water and CO_2 react, carbonic acid is produced and dissolves calcium and magnesium from carbonate rocks. It is generally accepted that $CaCO_3$ mg/L, 0-60 = soft, 61-120 = moderately hard, 121-180 = hard, and 181 and above = very hard water.

5.2.4.1 Estimation

Take suitable quantity of water (5-10 ml) and follow the procedure of Ca and Ca + Mg estimation as described for soil exchangeable Ca and Ca + Mg estimation (3.9).

Ca or Ca+ Mg, me/L = ml EDTA used x 0.01 X 1000/ ml of aliquot taken for analysis

5.2.5 Estimation of Sodium

Most of the laboratories nowadays use flame photometer for the determination of sodium in solution. Once set up and calibrated, the flame photometric method is rapid and reliable. In water sodium varies dramatically depending on the salinity level.

5.2.5.1 Reagents

Standard sodium solution: Dissolve 2.541 g of sodium chloride (NaCl, AR grade) in one litre volumetric flask in distilled water. This solution contains 1000 ppm Na in Standard Stock Sodium Solution.

Working sodium standard solution: Transfer 10 ml of 1000 ppm Sodium Standard Stock Solution in 100 ml of volumetric flask, dilute with water up to the mark and mixe well. This solution contains 100 ppm Na.

5.2.5.2 Procedure

1. Follow the procedure of the preparation of flame photometer for measurement as described for exchangeable (available) K measurement (3.8).
2. Transfer 2, 4, 6, 8, 10, 15 and 20 ml of working standard of 100 ppm Na solution to 100 ml of volumetric flasks and dilute up to the mark with distilled water. These flasks contain 2, 4, 6, 8, 10, 15 and 20 ppm Na in solution.
3. Select the sodium filter (589 nm).
4. Automise the flame photometer reading to distilled water. Thereafter, automise the prepared 2-20 ppm Na standard solution.
5. Prepare the standard curve by plotting concentration of Na vs. flame photometer reading.

6. Run the water samples in the similar way as for standard and record the reading.
7. Calculate the concentration of sodium from the standard curve.

me Na/L = ppm Na from standard curve/23 (i.e., equivalent weight of sodium)

5.2.6 Turbidimetric Method of Sulphur Determination

Traces of sulphate occur universally in all types of waters, its content may be appreciable in most saline waters showing EC greater than 1 dS/m (25°C). Most of the soil testing laboratories uses turbidimetric method for the determination of sulphate S in the extract. This method is based upon the principle of turbidity development on precipitation of suspended barium sulphate when excess $BaCl_2$ is added to a solution containing sulphate S in a slightly acid medium. Gum acacia, gelatine or glycerol is usually added to stabilize the suspension. The density of the suspension is read in a colorimeter or nephelometer. Although the turbidity method is rapid and convenient it is liable, as are all turbidimetric determinations, to give erroneous results if the soil contains much organic matter (Hesse, 1957). The interference of organic matter can be eliminated by adding activated charcoal (Bardsley and Lancaster, 1960).

5.2.6.1 Determination in solution (Combs *et al.* 1998)

5.2.6.1.1 Reagents

Barium chloride crystals: Ground $BaCl_2$ crystals in a mortar to pass a 0.5 mm sieve and be retained on a 0.20-30 mm sieve.

Seed solution (20 ppm S in 5.8 M HCl): Dissolve 0.1087 g of AR grade K_2SO_4 in 500 ml of distilled water and add 500 ml of concentrated HCl. Drop in a Teflon-coated magnet and place on a magnetic stirrer. Add 2 g of powdered gum acacia slowly while stirring to avoid formation of lumps. Keep refrigerated.

Standard sulphate solution: Dissolve 0.6793 g of AnalaR quality of potassium sulphate into a 500 ml volumetric flask in distilled water and make the volume up to mark. This is Stock Standard of 250 ppm SO_4^{-2}-S.

5.2.6.1.2 Preparation of standard curve

1. Pipette 10 ml 250 ppm Stock Standard of 250 ppm SO_4^{-2}-S into a 100 ml volumetric flask, dilute with distilled water to the mark, stoppered and mixe thoroughly. This solution contains 25 ppm SO_4^{-2}-S of Working Standard.
2. Transfer 2, 4, 6, 8, and 10 ml of Working Standard 25 ppm SO_4^{-2}-S into the 50 ml volumetric flasks, add 1 ml of 'seed solution' and dilute to 20 ml with distilled water.
3. Add approximately 1 g of the high-grade $BaCl_2$ crystals from a standard cup and mix the reagent thoroughly by inverting the flasks several times.
4. Make the volume 50 ml with distilled water. The flasks contain 1, 2, 3, 4 and 5 ppm concentration of SO_4^{-2}-S.
5. Again invert the flask several times and measure the turbidity (optical density/ transmission) with a spectrophotometer at 490 nm or with blue filter.
6. Set 100 % transmission or zero optical density with distilled water or extracting solution to which no reagents have been added.
7. Plot the percentage transmittance reading against concentration on semi-log graph paper.

5.2.6.2 Test Solution

Take 10 ml of water in 50 ml volumetric flasks and follow the procedure of standard curve preparation. Read the ppm sulphate-S in the water from standard curve.

5.2.7 Boron

Although boron is an essential micronutrient for plant growth and require smallest quantity. Boron toxicity has been a problem in connection with soils irrigated with water containing concentration of >2ppm, is harmful to most crops. Good quality of waters contains B below 1 ppm. Its concentration mainly depends upon its mineralogical source, nature of soil strata through which it percolates and pollution of water by boron-rich industrial effluents. Boron is usually high in saline ground water of the arid and semi arid regions. Boron toxicity occurs in soils fertilized with potash carrying a high boron content.

5.2.7.1 Determination

Boron can be estimated titrimetrically in macro and semimicro amounts, and colorimetrically or spectroscopically. The colorimetrically method are more convenient than the titrimetrically due to its easiness. The methods based on anthraquinone, such as the quinalizarin and carmine methods, all have their colours developed in sulphuric acid and therefore temperature change causes great variation in colorimetric readings (Jackson, 1973). Shanina *et al.* (1967) used azomethine-H as the reagent to form a coloured complex of boric acid in aqueous media and colour persists several hours without interference of a variety of salts. The method given here is described by Bingham (1982).

5.2.7.1.1 Apparatus

Spectrophotometer, polypropylene test tube, 10- ml capacity.

5.2.7.1.2 Reagents

Buffer solution: Dissolve 250 g of ammonium acetate and 15 g of EDTA disodium salt in 400 ml water. Slowly add 125 ml of glacial acetic acid, and mix.

Azomethine-H: Dissolve 0.45 g of azomethine-H in 100 ml of 1% L-ascorbic acid solution. Fresh reagent should be prepared weekly and stored in a refrigerator.

Standard solution of boron: Dissolve 0.114 g boric acid (H_3BO_3) in distilled water and bring to 1 litre volume with distilled water in volumetric flask. This solution contains 20 ppm B (stock solution). Dilute 2.5, 5.0, 7.5, 10.0, 15.0, 20.0 ml stock solution to 100 ml with distilled water in volumetric flasks. These solutions contain 0.5, 1.0, 1.5, 2.0, 3.0, and 4.0 ppm B, respectively.

5.2.7.1.3 Preparation of standard curve

Pipette 1 ml of each standard (0.5-4.0 ppm) into a 10 ml polypropylene tube, add 2 ml of buffer, and mix. Add 2 ml of azomethine-H solution, and mix well. Run one blank with 1 ml of distilled water, and proceed as for standard solutions. Read the absorbance of blank, standards, and samples after 30 minutes at 420-nm wavelength. Prepare the standard curve, plotting absorbance against the respective B concentrations.

5.2.7.2 Boron in samples

Take one ml of test solution (soil, plant or water), and proceed as for the preparations of standard. Read the B concentration in the unknown from the calibration curve.

B (ppm) = ppm B from calibration curve X A/Wt

Where: A = total volume of extract (ml)

Wt = weight of soil/plant

5.2.8 Chloride (Cl) by Mohr's titration method

Soil testing for Cl is usually done for measuring the salt component in saline soils. Chloride is not toxic to animals and humans, but some people can detect a salty taste at 250 mg/L. Water with Cl may also have a high sodium content. High chloride also may speed up corrosion in plumbing (just as road salt does to car). The concentration of chloride in drinking water below 10 mg/L is desirable.

Volumetric estimation of chloride in solution is based upon the titration of solution with silver nitrate using potassium dichromate as indicator (Mohr titration). In principle, the chloride in solution first forms a white precipitate of AgCl, then the colourless K_2CrO_4 indicator forms a brown-red precipitate of Ag_2CrO_4 that indicates the end of titration. The concentration Cl in solution can be obtained by calculating the volume and normality of standard $AgNO_3$ solution used.

5.2.8.1 Reagents

1. *Potassium chromate (K_2CrO_4) Indicator (5%) solution:* Dissolve 5 g of K_2CrO_4 in about 75 ml distilled water and add a saturated solution of $AgNO_3$ drop wise until a slight permanent red precipitate is formed. Filter and dilute to 100 ml.

2. *Standard silver nitrate ($AgNO_3$, 0.02 M):* Dissolve 3.398 g of $AgNO_3$ in distilled water and make the volume up to one litre. Standardize it against standard NaCl solution and keep in amber coloured bottle away from light.

5.2.8.2 Procedure

Take suitable aliquot of water sample in a 100 ml beaker and make it neutral or slightly alkaline with $NaHCO_3$. Titrate against the standard $AgNO_3$ (0.02 M) solution using 5-6 drops of K_2CrO_4 indicator (making it dark yellow) with continuous stirring till the first brick red tinge appears. A permanent reddish-brown precipitate indicates the end point. A blank titration containing suspension of calcium carbonate and indicator assists in matching colours.

5.2.8.3 Calculation

$$Cl\ (me/L) = \frac{\text{ml of 0.02 M } AgNO_3 \text{ used X 20}}{\text{Sample (ml) taken}}$$

5.2.9 *Nitrate (NO_3^-)*

The concentration of NO_3^--N rarely exceed 10 mg/L and are usually less than 1.0 mg/L. The higher levels are frequently associated with human contamination from fertilizers and sewage. High concentrations of nitrates can stimulate the growth of aquatic plants and may cause health hazard to juvenile mammals. In digestive system, nitrates are reduced to nitrites and enter the bloodstream. It causes the oxidation of haemoglobin to methemoglobin rendering it incapable of transporting oxygen. Cyanosis due to methemoglobinemia (blue baby disease) may result from drinking water with high nitrate concentrations. The maximum allowable level for NO_3^--N in drinking water has been set at 10.0 mg/L by the U.S. Public Health Service.

The nitrates in water can be determined with phenoldisulphonic acid method depends upon the nitration of position of 6 of 2, 4-phenoldisulphonic acid in fuming sulphuric acid:

$$C_6H_3OH(HSO_3)_2 + HNO_3 \rightarrow C_6H_2OH(HSO_3)_2\ NO_2 + H_2O$$

The product is colourless in acid and yellow when neutralized or in alkaline solution in the virtual absence of water. KOH or NH_4OH is therefore employed to shift the pH to yellow range for the colorimetric determination.

5.2.9.1 Reagents

Phenol 2, -4-Disulphonic acid: Dissolve 25.0 g of pure phenol in 150 ml of concentrated sulphuric acid in 500 ml beaker. Then add 75 ml fuming sulphuric acid. Mix the solution and boil for 2 hours. Keep the solution in brown bottle.

0.6% Ag_2SO_4 solution: w/v in water.

Nitrate free 6 N NH_4OH, $Ca(OH)_2$, $MgCO_3$.

Extraction solution: Mix 200 ml 1 N $CuSO_4$ solution and 1 litre of 0.6% Ag_2SO_4 solution and dilute to 10 liter with distilled water.

Standard nitrate solution: Dissolve pure dry 0.7221 g of KNO_3 in water and dilute to one litre. This prepared stock solution is 100 ppm NO_3^--N.

5.2.9.2 Preparation of standard curve

Dilute 10 ml of stock solution to 100 ml in a volumetric flask to give 10 ppm of NO_3^--N. Take 0, 2, 5, 10, and 15 ml of 10 ppm working solution in a separate porcelain evaporating dishes and evaporate to dryness on the steam bath in the atmosphere free from HNO_3. Cool the dishes and add 3 ml of phenoldisulphonic acid rapidly directly in the center of each. Rotate the dish to mix all the residual salt with reagent and leave for 10-minutes. Then add 15 ml water, and stir the solution with a glass rod until the all residues get dissolved. In the cool dish, add 6 N NH_4OH slowly until the solution is distinctly alkaline indicated by yellow colouration, add 3 ml extra. Transfer the aliquot of each dish separately in 100 ml volumetric flask with water completely and dilute to 100-ml. This series of dilution in 100 ml contains 0, 0.2, 0.5, 1, and 1.5 ppm NO_3^--N. Take either transmission of the nitrate solutions in a spectrophotometer with 420-nm light maximum. Prepare the standard curve by plotting the standard nitrates vs. transmission percentage on semilogrithmic paper to calculate the concentration of test solutions.

5.2.9.3 Water NO_3^-N

Take 100 ml water sample in the evaporating dish and add 0.25 g of $CaCO_3$ to make it alkaline. Evaporate to dryness, add 100-ml extraction solution and shake on rotary shaker for 10 minutes. Then add 0.2 g $Ca(OH)_2$ and continue shaking further for 5 minutes. Finally add 0.5 g of $MgCO_3$, shake the suspension and allow standing for a few minutes. Filter through dry filter paper. The first 10 ml may be discarded to rinse the collecting glasswares and about 40 ml to be collected. If filtrate is coloured with organic matter, 1 g nitrate-free activated charcoal is added to the suspension, followed by shaking and filtration.

Take 20 ml aliquot of the clear filtrate in the evaporating dish and analyze the NO_3^--N as described for standard curve preparation. The concentration of NO_3^--N in water as ppm N is obtained from the standard curve.

$$\text{ppm } NO_3^- N \text{ in water} = \frac{\text{ppmNfromstandardcurve x (ml) final colour volume}}{\text{(ml) extraction solution x (ml) water evaporated}}$$

5.2.10 Phosphorus

The Murphy and Riley (1962) method suggested for the determination of orthophosphate in water is not sensitive enough for low concentration of phosphates. The malachite green method developed by Ohno and Zibilske (1991) is very much sensitive to determine the P in natural waters and limited volume of extracts.

5.2.10.1 Reagents

Ammonium molybdate solution: Prepare 14.2-mmol/L ammonium molybdate tetrahydrate in 3.14 mol/L H_2SO_4.

Polyvinyl alcohol: Heat 400 ml water in a beaker to 80°C, add 3.15 g polyvinyl alcohol (PVA) and stir until dissolved. Cool the solution to room temperature and add 0.35 g malachite green in the solution. Transfer the prepared solution to one litre volumetric flak and make the volume with water.

Working standard phosphorus solution (0.50 ppm P): Transfer 5 ml of 50 ppm P stock solution to 500 ml volumetric flask and dilute with water up to the mark and mix well. This standard solution contains 0.50 μg of P.

5.2.10.2 Preparation of standard P curve

Transfer 0.5, 1.0, 2.5, 5.0, 7.5, 10.0, 15.0, 20.0, and 25.0 ml of working standard 0.5-ppm P solution into 50 ml volumetric flaks. Dilute solution of flasks to around 40 ml with water. Add 2 ml of ammonium molybdate solution and mix. After 10 minute, 2 ml polyvinyl alcohol is to be added and mixed thoroughly after making the volume with water. This prepared final standard solutions contain 0.005, 0.01, 0.025, 0.05, 0.075, 0.10, 0.15, 0.20, and 0.25 ppm P. Measure the absorbance after 30 minutes of colour development at 630 nm with spectrophotometer/ colorimeter. Prepare the standard curve by plotting absorbance against the concentration of P. Prepare one control without P solution (only distilled water) and follow the similar procedure as for standard curve to check impurity of chemicals used.

5.2.10.3 Water phosphorus

Filter the water sample through a membrane filter (0.45-μm pore diameter). Take 25.0 ml of sample in 50 ml volumetric flask. Proceed for the development of colour for the estimation of phosphorus as described for the standard curve preparation.

$$\text{ppm P in water (mg P/L)} = \text{P concentration from curve} \times 50/25$$

5.2.11 Quality of Irrigation Water

The quality of irrigation waters is dependent on the total salt content, on the nature of the salts (particularly of Na and B) present in solution, and proportions of me of Na to me of other metallic cations, and of Ca and Mg to bicarbonates. The quality of water can be evaluated in terms of the conductance, sodium adsorption ratio (SAR) and residual sodium carbonate (RSC). The sodium adsorption ratio (SAR) is determined by using the equation:

$$SAR = \frac{[Na]}{\sqrt{[Ca + Mg/2]}}$$

Where all concentrations are expressed in me/L.

The "residual Na_2CO_3" (RSC) of water is defined as follows (Jackson, 1973):

Residual Na_2CO_3 = (CO_3^{2-} + HCO_3^-) - (Ca^{2+} + Mg^{2+})

Where all the quantities are expressed as me/L.

Irrigation water with poor quality may cause salinity, specific ion toxicity or infiltration problems in soils. Such effects of irrigation may adversely affect crop production. Gupta *et al.* (1994) mapped the ground waters quality for irrigation into saline and alkaline waters for the country on a large scale (1:6 million scale) into the four following categories:

1. Good water, EC < 2 and SAR<10
2. Saline water, EC >2 and SAR <10
3. High-SAR saline, EC >4.0 and SAR >10
4. Alkaline water, EC variable; SAR variable and RSC >2.5

Such type of problems is mainly restricted to the arid and the semi-arid regions of the country. Other than salinity/alkalinity constraints, water bodies may also face local specific problems due to excessive amounts of nitrates, fluoride, boron, magnesium or toxicity of metal caused by pollution. Irrigation with waters of SAR and high residual alkalinity may deteriorate soils conditions similar to those met within sodic soils. Long-term use of high alkaline waters induces more harmful in reducing the infiltration rates of the soils. The farmers of the arid and semi-arid regions continue to use waters, which would otherwise be termed unsuitable on the basis of criteria followed by the US Salinity Laboratory (Richards, 1954). In the light of the new experimental information emerged from the All-India Coordinated Research Project on Use of Saline Water, highly conservative standards have been replaced with the more realistic guidelines for monsoon agriculture (Table 11).

Table 11. Guidelines for using poor quality irrigation waters (Gupta and Manchanda, 1996).

Soil texture (% Clay)	Crop tolerance	Upper limits of EC (dS/m-1) in rainfall region (mm)		
		<350	350-550	550-750
Fine (>30)	S	1.0	1.0	1.5
	ST	1.5	2.0	3.0
	T	2.0	3.0	4.5
Moderately fine (20-30)	S	1.5	2.0	2.5
	ST	2.0	3.0	4.5
	T	4.0	6.0	8.0
Moderately coarse (10-20)	S	2.0	2.5	3.0
	ST	4.0	6.0	8.0
	T	6.0	8.0	10.0
Coarse (<10)	S	-	3.0	3.0
	ST	6.0	7.5	9.0
	T	8.0	10.0	12.5

S, ST and T indicate sensitive, semi-tolerant and tolerant crops to salinity of irrigation waters, respectively.

References

Bardsley, C.E. and Lancaster, J.D. (1960). Determination of reserve sulphur and soluble sulphates in soils. *Soil Sci. Soc. Am. Proc.* 24:265-268.

Bingham, F.T. (1982). Boron In: "*Method of Soil Analysis, Part 2*" (ed. Page, A.L.), pp. 431-447. *Am. Soc. Agron.* Inc. and *Soil Sci. Soc. Am. Inc.*, Madison, Wisconsin, USA.

Combs, S.M., Denning, J.L. and Frank, K.D. (1998). Sulphate-sulphur. In: *Recommended Chemical Soil Test Procedures for the North Central Region*"(ed. Brown, J.R.). North Central Regional Research Publication No. 221 (Revised), USA.

Gupta, R. K. and Manchanda, H.R. (1996). Management of poor quality waters. In: "*Soil Management in Relation to Land Degradation and Environment*" (Eds. Biswas, T.D. and Narayanaswami, G.), pp. 148-163.

Gupta, R. K, Singh, N.T. and Sethi, M. (1994). Ground Water Quality for irrigation in India (Map 1:6 million scale), Central Soil Salinity Research Institute Publication 19.

Hesse, P.R. (1957). The effect of colloidal organic matter on the precipitation of barium sulphate and a modified method for determining soluble sulphate in soils. *Analyst* 82:710-712.

Jackson, M.L. (1973). *Soil Chemical Analysis* (Indian ed.). Prentice Hale of India, Pvt. Ltd. New Delhi.

Murphy, J. and Riley, J.P. (1962). A modified single solution method for determination of phosphate in natural waters. *Anal. Chim. Acta.* 27:31-36.

Richards, L.A. (1954). *Agriculture Handbook*, US Department of Agriculture, 60.

Shanina, T.M., Gelman, N.E. and Mikhailovskaya, V.S. (1967). Quantitative analysis of heterorganic compounds. Spectrophotometric microdetermination of boron. *J. Anal. Chem.* USSR (Engl. Transl.) V. 22: 663-667. (Transl. From *Zh. Anal. Khim* 22:782-787.)

Ohno, T. and Zibilske, L. M. (1991). Determination of low concentrations of phosphorus in soil extracts using malachite green. *Soil Sci. Soc. Am. J.* 55:892-895.

—•—

6

Pesticide Residue Analysis in/on Plants, Water and Soils

India is recovering from the euphoria of the Green Revolution and battling the residual effects of the extensively used chemical fertilizers and pesticides in the country's soils. The decade from 1980 to 1990 alone saw the area under pesticides in India, increased a whopping 20-fold, from six million hectares to 125 million hectares. After a high annual consumption of nearly 75000 MT reached in the early '90s, interventions in the form of Integrated Pest Management (IPM) practices have only now started to show a declining trend in the use of pesticides in India. Interestingly, India's consumption of pesticides per hectare is low (0.288 kg/ha) when compared with world averages—0.5 kg/ha against Korea's 6.60 kg/ha and Japan's 12.0 kg/ha. According to the pesticides industry statistics, India spends US $3/ha on pesticides compared with US $24/ha spent by Philippines, US $255/ha spent by South Korea and US $633/ha by Japan. Yet, despite a comparatively low use of pesticides in India, the contamination of food products in the country is alarming. About 20 per cent of Indian food products contain pesticide residues above tolerance level compared to only 2 per cent globally. No detectable residues are found in only 49 per cent Indian food products compared to 80 per cent globally. A pesticide is any substance or mixture of substances intended for preventing, destroying, repelling or mitigating any pest. Though often misunderstood to refer only to insecticides, the term pesticide also applies to herbicides, fungicides and various other substances used to control pests.

Pesticides and their metabolites have received particular attention in the last few years in environmental trace organic analysis because they are regularly detected in food products, surface/ground waters and soil strata throughout the world as consequence of their widespread use for agricultural and nonagricultural purposes. So it is the need of the hour to study the levels of pesticides residues in plants, water and soil and their fate in environment after its degradation.

It would be necessary to work out tolerance limits of pesticides residues in/on crops, soil and water and the efforts, therefore, are directed towards various residue analytical methods like TLC, GLC and HPLC etc. which are suitable for the purpose.

6.1 Steps in Pesticide Residue Analysis

1. Sampling
2. Extraction
3. Clean up
4. Concentration (Evaporation)
5. Analysis

6.1.1 Sampling

The sample collected for analysis should possess accuracy, validity, representativeness and variability. Accuracy depends upon the original field from where the sample is collected. A valid sample is one which, when selected, ensures that each unit of the material in the population being sampled has an equal chance of being selected for test. For sample collection each field plot should be replicated at least three times. Atleast five sub samples from different parts of a plot should be collected and pooled together to form a sample per plot. The laboratory sample should be prepared from the field sample as rapidly after harvesting as possible in order to minimize residue loss, water content etc.

In case of soil samples, quartering is done by spreading the gross sample after thorough mixing in a tray. One quarter of sample is taken and three quarters are destroyed. The representative field sample may be placed in a clean polythene bag after proper labeling. According to Lykken (1963) the optimum amount of each sample supplied to laboratory for analysis should be approximately 1 kg. If it is to be analyzed later on, it should be kept in a deep freezer as soon as possible. It is always desirable to minimize storage period because there are reports of loss of some insecticides in storage.

Water samples are collected from one location at one particular time. Water samples can also be collected from different location at a site and are mixed to form a composite sample. Sometimes water samples are collected over a given period of time by an automatic sampling device. Collected water samples are chlorinated with sodium thiosulfate to remove any free chlorine and often use H_2SO_4 or HCl to adjust the pH (<2).

6.1.2 Extraction

The residues of insecticides are present in minute quantities in substrate. The next step in residue analysis is, therefore to transfer this very small amount from a relatively large proportion of biological material, on or in which it is found into a solvent. The method of extraction and the type of solvent combinations will be dependent on the chemical and physical properties of the pesticide to be extracted, the type of substrate from which it will be quantitatively removed and the final method of analysis. The methods employed to extract pesticides (Fig.3) from water, soils and plants are as follows;

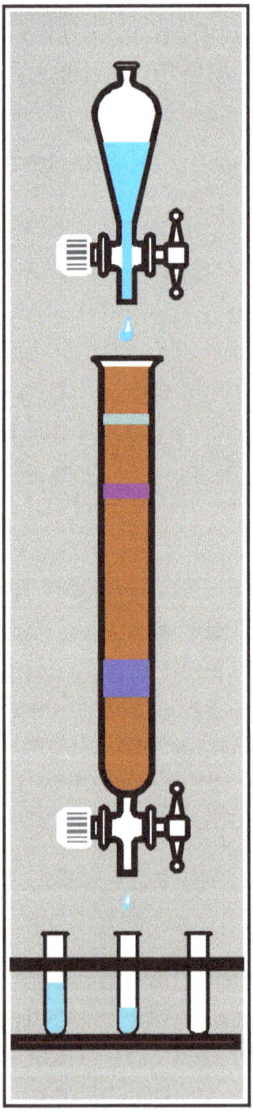

Fig 3. Extraction

6.1.2.1 Water

6.1.2.1.1 Liquid /Liquid extraction (EPA's official Method)

Insecticides generally have chemical structures foreign to the types of compound present in the substrate; hence in effect they also possess some physical and chemical properties unique to the intermediate environment. This distinctive nature of a pesticide facilitates its separation by favorable partition into one of two suitable immiscible solvents, with the substrate extractives largely favoring the other solvent. Naturally, such manipulation requires that the insecticide should dissolve the substrate extractives. Additional requirement is that the solvent in which the insecticide is preferentially dissolved should be low boiling for subsequent ease of removal.

6.1.2.1.2 Solid phase extraction (SPE)

Solid phase extraction (SPE) is an increasingly useful sample preparation technique. With SPE, many of the problems associated with liquid/liquid extraction can be prevented, such as incomplete phase separations, less-than-quantitative recoveries, use of expensive, breakable speciality glassware, and disposal of large quantities of organic solvents. SPE is more efficient than liquid/liquid extraction, yields quantitative extractions that are easy to perform, is rapid, and can be automated. Solvent use and lab time are reduced. SPE is used most often to prepare liquid samples and extract semi volatile or nonvolatile analytes, but also can be used with solids that are pre-extracted into solvents. SPE products are excellent for sample extraction, concentration, and cleanup. They are available in a wide variety of chemistries, adsorbents, and sizes. Selecting the most suitable product for each application and sample is important.

In Reversed phase SPE, silicas and nonpolar adsorption media usually are conditioned with a water-miscible organic solvent such as methanol, followed by water or an aqueous buffer. Methanol wets the surface of the sorbent and penetrates bonded alkyl phases, allowing water to wet the silica surface efficiently. Sometimes a pre-conditioning solvent is used before the methanol step. This solvent is usually the same as the elution solvent, and is used to remove any impurities on the SPE tube that could interfere with the analysis, and may be soluble only in a strong elution solvent.

6.1.2.2 Soil (Soxhlet Extraction, classical for solids)

This is the classical method for dry samples. Here the samples are dried, grinded, subdivided and extracted with suitable non-polar solvent. Anhydrous sodium sulphate is mixed with sample and then kept in porous thimble of Soxhlet apparatus and extraction requires 2-3 hours. This method is not useful for thermally unstable insecticides.

6.1.2.3 Plants

6.1.2.3.1 Surface rinsing

The extraction of the insecticides from the surface of the substrate is possible by this method. The sample is placed in the glass container to which solvent is added at a definite ratio, preferably the sample is completely dipped in such glass containers possessing sample and the solvent are shaken vigorously by hand or rotated on the electrical shakers for a definite time period preferably for one hour. This method is suitable to extract the surface residue from the plants and not the absorbed residue from plant tissues.

6.1.2.3.2 Blending

The extraction of most of the insecticides in plants is possible by blending or macerating them in a Waring Blender (Mixi) with solvent for some time and then the solvent possessing residues of insecticides is decanted through funnel containing Whatman Filter Paper No. 1 or filtered under pressure if emulsion in extractive is formed.

Anhydrous sodium sulphate over the funnel during filtration for entrapping water is to be added in filtered solvent extractive @ 0.1 g per 100 ml. In this method there is an intimate mixture of solid and solvent which results into the removal of external as well as internal residues.

6.1.2.3.3 Soxhlet extraction

It is useful for extraction of dry samples. The dried samples are often grinded and then extracted with suitable non-polar solvent in a Soxhlet Apparatus for a quite some time. It is not useful for extraction of thermo unstable insecticides. However, this is giving highest recovery from fortified samples and aged samples as compared to other methods.

6.2 Clean-up

Isolation of insecticides from interfering extraneous solvents is known as clean up. According to *Gunther* and Blinns (1955) the cleanup procedure is required under the following conditions;

(i) When the insecticide is evaluated by direct measurement.

(ii) When the insecticide can be measured only after all the interfering substances are removed or converted into products which shall not interfere during analysis.

(iii) When insecticide can not be measured in isolated form and is converted to a suitable derivative for measurement.

The amount of extract cleanup required prior to the final determination depends on the concentration of interferences in the sample, the selectivity of both the extraction procedure and the determinative method and the required detection limit. The following cleanup methods are generally used.

6.2.1 Steam Distillation

Many natural oils and fats are to some extent steam volatile, thus offering a means of separating them from non-volatile insecticides. Steam volatile insecticides, which form non-volatile salts or complexes can also be treated in this manner. Converse procedure is the steam distillation of intact insecticides. Aldrin, chlordane, heptachlor and lindane are readily steam volatile.

6.2.2 Oxidation

Most of the naturally occurring compounds in plants are oxidisable. One highly successful methods of determining organophosphorus pesticides is to oxidize the purified extract to inorganic phosphate with a mixture of fuming nitric acid and perchloric acid. The inorganic phosphate is then determined by molybdenum blue method.

6.2.3 Saponification

Fats, oils and other high triglycerides containing pesticide residue samples can be effectively cleaned up by saponification by refluxing the sample with alcoholic potassium hydroxide. The triglycerides are converted to soaps and glycerine. The soap solution is then extracted with n-hexane; with the pesticide partitioning into hydrocarbon phase while the soaps remain in the aqueous layer. Schafer *et al.* (1963) have used saponification to hydrolyse milk-fats and simultaneously dehydrochlorinate DDT to DDE which may be measured by GLC.

6.2.4 Partition

Extracts containing significant quantities of fats can be partitioned between immiscible polar solvent. Jones and Riddich (1952) found that pigments and waxes were preferentially soluble in n-hexane while pesticides because of their high polarities were extracted into the acetonitrile.

The acetonitrile can be evaporated or diluted with water and re-extracted with hexane in order to recover pesticides. N, N-dimethyl formamide and dimethyl sulfoxide can also be used to separate pesticides from liquids in place of acetonitrile (Edelman).

6.2.5 Sweep co-distillation

The sweep co-distillation is a useful method for clean-up of pesticides because it eliminates the need for specialized adsorbents and equipments and large volume of costly solvents. In sweep co-distillation, the substrate extractive is injected into a glass column packed with glass wool. This is followed by repeated injections of the solvent. Nitrogen carrier gas sweeps the vapourised volatile compounds including the pesticides through the column and the interfering organic substances remain on the packing material while the pesticides are collected for analysis (Gunther *et al.*, 1966).

6.2.6 Coagulation

This method is used for clean-up of carbamates. An aqueous solution of ammonium chloride and phosphoric acid is added to the filtered solution to precipitate the interfering substances which are then removed by filteration. After filteration the insecticide is partitioned in a water immiscible solvent. Benson and Finochiero (1965) applied the coagulation method for clean-up of carbaryl residues.

6.2.7 Adsorption chromatography

In pesticide residue analysis, selective adsorption of the interferences on an active solid is the procedure most commonly followed. The active solids most commonly used are florisil, activated carbon, activated alumina, celite and magnesium oxide. Polyethylene coated alumina is probably effective in both adsorptive and absorptive processes for removal of liquids and other interfering substances. In some cases, a simple adsorbent is not capable of removing all the interfering materials in an extract and consequently mixtures of different absorbents are often employed.

6.3 Concentration (Evaporation)

6.3.1 Kuderna-Danish concentrators

Evaporation in K-D is most suitable for solvents with relatively low boiling points, such as ethyl ether (b.p. 34.6° C), methylene chloride (40.5°C), and acetone (56.5°C), or for solvent mixtures that form low boiling azeotropes. Analytes must be able to withstand steam bath temperatures for concentration by K-D.

6.3.2 Rotary evaporator

Vacuum and heat combine to reduce the quantity of solvent in a rotary evaporator. Solution in a round-bottom (r-b) flask is simultaneously rotated and heated by a water bath, while a vacuum pulled from the end of a condenser increases the rate of evaporation. Solvent evaporated from the solution is collected in a second flask attached to condenser.

Evaporation of solvent using rotary evaporation is recommended for heat-labile residues, because temperature of the water bath used is much lower than that of a steam bath. Vacuum withdrawal of vapours, combined with water bath temperature of about 30°C, adequately removes solvents such as methylene chloride (b.p. 40.5°C) without damaging analytes. Evaporation in a rotary evaporator is more rapid than in a K-D, but is limited to only one solution at a time.

6.4 Analysis

The estimation of residues is possible by adopting any one of the analytical techniques viz., microbioassay, enzymatic methods, spectrophotometric assay, colorimetric method, chromatographic method, radiometric method, etc. Amongst these residue estimation methods, following are important.

6.4.1 Enzymatic methods

This method is suitable to detect the residues of those insecticides which inhibit the cholinesterase enzyme (ChE). Organophosphate and carbamate and their metabolites are strong inhibitors of cholinesterase enzyme, therefore, the residues of them can be determined by this procedure.

During analysis the inhibitor (insecticide residue) is allowed to react with cholinesterase enzyme resulting in the inhibition of certain amount of cholinesterase after an incubation period. After addition of acetylcholine, the inhibited cholinesterase hydrolyses the acetylcholine to give acetic acid, chloline and unhydrolysed acetylcholine. The greater the amount of uninhibited cholinesterase, the greater will be the acetic acid and the amount of unhydrolysed acetylcholine will be lesser.

Different methods for assessment of cholinesterase activity are utilized for determination of residues of organophosphate and carbamate insecticides.

(a) Potentiometric method

(b) Titrimetric method

(c) Manometric method

(d) Colorimetric method

(e) Thin Layer Chromatography method

6.4.1 Potentiometric Method

The enzyme cholinesterase is allowed to act on acetylcholine in a buffer solution for a defined period of time, usually an hour at a constant temperature. The pH of the mixture is measured both at the beginning and the end of the reaction. The change in pH due to release of acetic acid is a measure of the enzyme activity. Since this method has been adopted widely in the country stepwise procedure for detection of residues is given as under.

(i) Take 1 ml insecticidal solution or sample extract into 10 ml Griffen beakers, control should be run simultaneously.

(ii) To check loss of insecticide from solution, add 0.5 ml glycerol solution to each beaker. Evaporate methanol with a gentle stream of cool air.

(iii) Add 3 ml of cholinesterase solution to such beaker at intervals of 2 minutes.

(iv) Put 2 small round glass beads in each beaker to facilitate agitation during shaking.

(v) Shake the contents of the beaker on an incubator shaker at a constant temperature of 25 ± 1 °C.

(vi) After 30 minutes, the initial pH of the content should be recorded.

(vii) At exactly 32 minutes, 1 ml acetylcholine bromide should be added to the beakers and the reaction should be allowed to go for 60 minutes and therefore the final pH must be recorded.

(viii) The percent inhibition of enzyme can be worked out.

$$\text{Enzyme inhibition (\%)} = \left[1 - \frac{\Delta\,\text{pH of sample}}{\Delta\,\text{pH of control}}\right] \times 100$$

Δ pH = pH initial – pH final

Reagent Preparation

- *Glycerol solution:* Take 10 ml glycerol in 100 ml volumetric flask and add 9 ml absolute methanol up to the mark.
- *Cholinesterase solution:* Prepare fresh by mixing 9.6 ml blood plasma (human or horse), 10.4 ml saline solution and 40 ml barbital buffer solution and keep cool in ice bath.
- *Saline solution:* Dissolve 9 g sodium chloride in one litre distilled water
- *Barbital buffer solution:* Dissolve 7.42 g of sodium barbital, 89.46 g of potassium chloride and 1.09 g of monobasic potassium phosphate in one litre glass distilled water. Adjust pH to 8 with 0.1 N hydrochloric acid.
- *Acetylcholine bromide:* Dissolve 4.97 g of acetylcholine bromide in 100 ml distilled water.

6.4.1.2 Titrimetric method

The principle of titrimetric method is the determination of the liberated acetic acid by titration with NaOH/KOH to a constant pH. Cresol red used as indicator for acetylcholine hydrolysis.

6.4.1.3 Manometric method

The Warburg manometric cholinesterase method is based on the amount of carbon dioxide liberated from sodium bicarbonate added to the reaction medium. The results may be expressed as microlitres of CO_2 evolved from the bicarbonate buffer by the acetic acid produced during the reaction.

6.4.1.4 Colorimetric method

On the completion of enzymatic reaction unchanged acetylcholine is determined colorimetrically. In the presence of excess ferric ion the purple colour of the ferric acethydroxamate (acetylcholine + hydroxylamine) is measured at 540 mm.

6.4.1.5 TLC method

The principle involved in TLC enzyme inhibition method is based on the equation given below.

(i) Enzyme + substrates ↔ Enzyme substrate → enzyme + product

(ii) Enzyme + substrate ↔ Enzyme inhibited + substrate

The reaction taking place can be detected on TLC by using pH indicators of chromogenic substances.

6.4.1.6 Spectrophotometric method

The colorimetric method of analysis includes the treatment of extract of sample substrate in sodium with the reagents that will cause the development of colour. The term colorimeter is usually restricted to an instrument where the eye is used to judge colour; a photometer replaces the eye with a photoelectric cell, while a spectrophotometer employs some sort of monochromator to isolate light or radiation of a particular wavelength for an analysis. The radiant energy is divided into three regions *viz.,*

Ultraviolet : 185 –400 mμ

Visible : 400-800 mμ and

Infrared : 800 mμ - 1600 mμ and beyond.

Three types of spectrophotometry *ie.,* ultraviolet spectrophotometry, visible photometry and infrared photometry based on above referred regions of radiant energy are utilized for detection of residues of insecticides.

The regression equation *i.e.* $y = a + bx$ is first prepared by analyzing the standard doses (x) of the insecticide spectrophotometrically and the absorbances (y) recorded against them at a particular wavelength. Through the regression equation, the calibration curve (reference curve) is prepared. The absorbance recorded from the sample extracts on spectrophotometer are set in the regression equation or compared with the calibration curve for computing the amount of residue present.

Compounds which are capable of absorbing electromagnetic radiation and then at later stage emitting atleast a portion of the absorbed radiant energy as visible light are said to be fluorescent or *phosphorescent*. After absorption the compound emit light for micro seconds are known as fluoresce where as phosphorescent chemicals need longer periods of time before light energy is emitted. Based on these Photoflurometric and Phosphorometric methods are developed for some insecticides. These are highly sensitive analytical tool.

6.4.1.7 Chromatographic method

Chromatography is a separation process based on the distribution between a solid and mobile phase. Different types of *Chromatography* are;

a) Paper Chromatography (PC)
b) Thin layer Chromatography (TLC)
c) Column Chromatography (CC)
d) Gas liquid Chromatography (GLC)
e) High performance Liquid Chromatography (HPLC)

6.4.1.7.1 Paper chromatography (PC)

Paper chromatography is a sub-division of solution (liquid) chromatography based on the nature of the sortive system. The stationary phase is a paper sheet containing water or some other polar phase and the solvent is usually organic in nature.

6.4.1.7.2 Thin layer chromatography (TLC)

Cellulose or thin silica gel or an aluminium oxide layer on a thin plate made out of glass, polyester, or aluminium are used in TLC. A sample spot is applied on the start line. The plate is set into a closed chamber with a small amount of solvent (sample spot facing the bottom). The solvent flows from bottom to top on the plate (Fig. 4). This method is simple, cost efficient and universal and is useful for analysis of small quantity of sample.

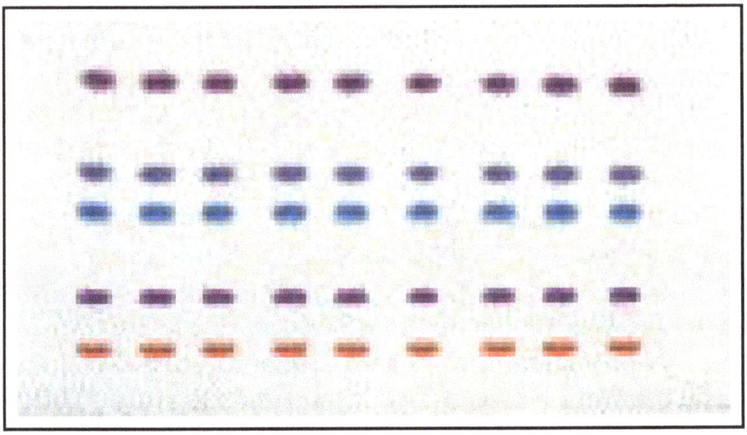

Fig.4. TLC

6.4.1.7.3 Column chromatography

In column chromatography (Fig.5), the stationary phase, a solid adsorbent, is placed in a vertical glass (usually) column and the mobile phase, a liquid, is added to the top and flows down through the column (by either gravity or external pressure). Column chromatography is generally used as a purification technique: it isolates desired compounds from a mixture.

The mixture to be analyzed by column chromatrography is applied to the top of the column. The liquid solvent (the eluent) is passed through the column by gravity or by the application of air pressure. An equilibrium is established between the solute adsorbed on the adsorbent and the eluting solvent flowing down through the column. Because the different components in the mixture have different interactions with the stationary and mobile phases, they will be carried along with the mobile phase to varying degrees and a separation will be achieved. The individual components, or elutants, are collected as the solvent drips from the bottom of the column.

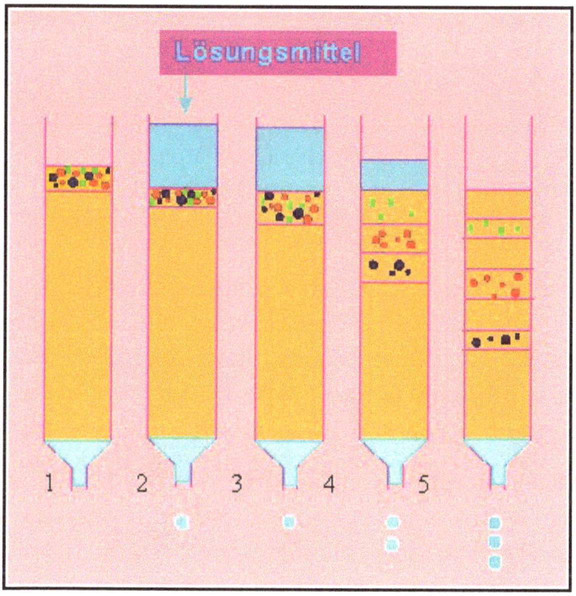

Fig .5. Column chromatography

6.4.1.7.4 Gas chromatography

Gas chromatography is mostly used for both qualitative and quantitative analysis of residues of insecticides (Fig. 6). It can separate insecticides and their metabolites from the mixture of insecticides. This method is very sensitive and can detect the residues as low as nano or picogram level. The selectivity and sensitivity of the technique is constantly being increased by improved resolution of the components with improved column packings and with the design of more

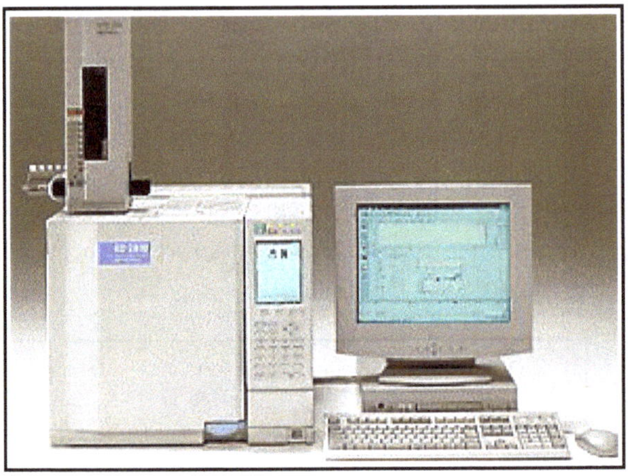

Fig. 6. Gas chromatography

sensitive detectors. Gas chromatography is the method of separation of vapourized components which are distributed between a fixed stationary phase and a moving inert gas phase. It is of two types. Gas solid chromatography (GSC) where the gas is the moving phase and solid is a stationary and Gas liquid chromatography (GLC) where the stationary phase is a non-volatile liquid distributed on a solid support.

Basic gas chromatographic apparatus consists of (a) source of gas- a gas cylinder fitted with pressure regulator, (b) drying cartridge, (c) inlet. Sample in the from of solution is introduced by injection with microsyringe (1-10μl capacity); the inlet portion is heated to vapourize the sample instantly before it reaches column, (d) column- the vapours of samples are passed through column which is U shaped or coiled tube made up of glass, aluminium or copper. The columns contain the stationary phase, generally a partitioning liquid coated on some inert material, (e) detector- after the sample is separated on column it enters the preheated detector which senses the emergence of individual component. The detectors can be microcolorimetric, electron-capture, flame photometric, thermionic etc., (f) amplifier- the signal from the detector is amplified several hundred folds before feeding to a record, (g) recorder- amplified signal is recorded as peaks called chromatograms.

The versatility of gas chromatography is dependent upon the type of columns used and their packings and the type of detectors that can be utilized for separation, identification and quantification of compounds.

Peak height and area of the recorded chromatography curves are related to the amount of insecticide. Peak area is usually a more reliable measurement than peak height. Area measurement is usually done with a planimeter, a disc integrator attached to the recorder pen. During standardization, series of peak areas on different concentration of insecticide are used for determining the standard calibration curve. The relation of concentration to peak area with the optimum operation is usually forming a straight line or if there is slight variation, this can be subjected for regression statistical analysis.

It is the method of separation of vapourized components which are distributed between a fixed stationary phases and a moving inert gas phase. It is of two types (i) GC and (ii) GLC. Very long (up to 200 m) very thin (0.1-0.5 mm) capillaries out of fused silica (silica or quartz). The inner layer of capillaries is coated with a thin layer (solid phase). The mobile phase is always a gas and serves as a carrier (helium, nitrogen, or hydrogen). The (vapourizable) sample is added over an injector. Gas chromatography involves different detectors, it is a quick method with high separation efficiency but complicated as only small vapourizable sample can be analyzed.

6.4.1.7.5 High performance liquid chromatography (HPLC)

It is a modification of Column Chromatography (Fig.7). Silica gel, modified silica gel or polymer form the solid phase that shows a high separation efficiency due to the small particle size. However, a high pressure is needed and therefore, the material is filled into stainless steel columns. The sample is injected over the injection valve and then the solvent is pumped through the column at high pressure (up to 400 bar). Numerous detectors such as: UV, RI, Fluorescence, etc. can be used. It is suitable for small and large amounts of substance, high separation efficiency and automation.

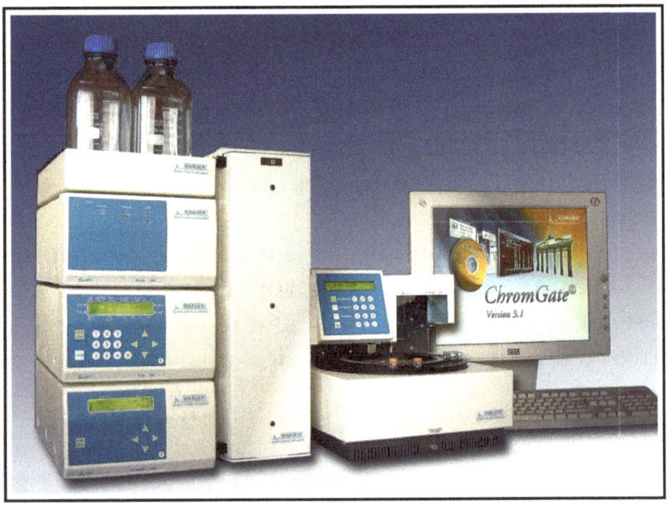

Fig.7. HPLC

HPLC is used for qualitative and quantitative analysis. In qualitative analysis, by comparing retention times (RT) we can identify the compounds and in quantitative analysis it is the process of determining the unknown concentration of a compound in a known solution.

6.4.1.7.5.1 Modes of operation

On the basis of type of column material packed (Polar or Non-polar) two types of modes of operation are present in HPLC, out of this Reversed Phase (RP) HPLC, has 90% of application in environmental analysis because of its wider adaptability.

6.4.1.7.5.1.1 Normal phase (NP) HPLC

The solid phase consists of silica gel or polar modified silica gel. The mobile phase consists of heptane, hexane, ethyl acetate, Isopropanol

6.4.1.7.5.1.2 Reversed phase (RP) HPLC

The solid phase consists of non-polar modified silica gel or polymer. Mobile phase consists of Isopropanol, methanol, water, and acetonitrile. A typical phase is C18 (carbon chain with 18 C-atoms).

On the basis of use of HPLC there are of two types:

1. Analytical HPLC focuses to obtain information about the sample compound. The information that can be obtained includes identification, quantification, and resolution of a compound. It has columns with a small diameter (2-8 mm) and a very small particle size (3-5μm) to identify and quantify substances (small substance amounts up to trace analysis)
2. Preparative HPLC refers to the process of isolation and purification of compounds. Important is the degree of solute purity and the throughput, which is the amount of compound, produced per unit time. It has columns with a large diameter (>8 mm) and a larger particle size (35 μm) to separate and purify substances.

6.4.1.7.5.2 Components of HPLC

6.4.1.7.5.2 .1 HPLC pumps

There are several types of pumps available for use with HPLC analysis, they are Reciprocating Piston Pumps, Syringe Type Pumps, and Constant Pressure Pumps. Reciprocating Piston Pumps are most common and consist of a small motor-driven piston which moves rapidly back and forth in a hydraulic chamber that may vary from 35-400 μL in volume. On the back stroke, the separation column valve is closed, and the piston pulls in solvent from the mobile phase reservoir. On the forward stroke, the pump pushes solvent out to the column from the reservoir. A wide range of flow rates can be attained by altering the piston stroke volume during each cycle, or by altering the stroke frequency. Dual and triple head pumps consist of identical piston-chamber units, which operate at 180 or 120 degrees out of phase. This type of pump system is significantly smoother because one pump is filling while the other is in the delivery cycle.

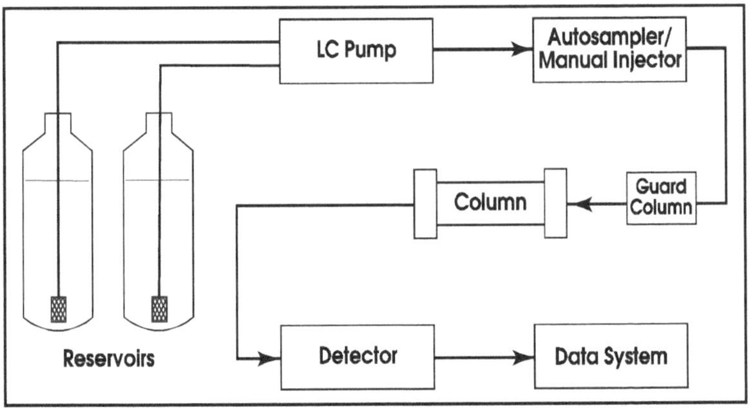

Fig 8. HPLC System

6.4.1.7.5.2.2 Injectors for HPLC

Samples are injected into the HPLC via an injection port. The injection port of an HPLC commonly consists of an injection valve and the sample loop. The sample is typically dissolved in the mobile phase before injection into the sample loop. The sample is then drawn into a syringe and injected into the loop via the injection valve. A rotation of the valve rotor closes the valve and opens the loop in order to inject the sample into the stream of the mobile phase. Loop volumes can range between 10 μL to over 500 μL. In modern HPLC systems, the sample injection is typically automated.

6.4.1.7.5.2.3 Columns

Guard Columns are placed anterior to the separating column. This serves as a protective factor that prolongs the life and usefulness of the separation column. They are dependable columns designed to filter or remove: 1) particles that clog the separation column, 2) compounds and ions that could ultimately cause "baseline drift", decreased resolution, decreased sensitivity, and create false peaks, 3) compounds that may cause precipitation upon contact with the stationary or mobile phase, and 4) compounds that might co-elute and cause extraneous peaks and interfere with detection and/or quantification.

Capillary columns are also known as microcolumns and have a diameter much less than a millimeter and these are three types: open-tubular, partially packed and tightly packed. They allow the user to work with nanoliter sample volumes, decreased flow rate and decreased solvent volume usage which may lead to cost effectiveness.

Fast columns are to obtain improved sample throughout (amount of compound per unit time). For many columns, increasing the flow or migration rate through the stationary phase will adversely affect the resolution and separation. Therefore, fast columns are designed to decrease time of the chromatographic analysis without forsaking significant deviations in results. These columns have the same internal diameter but much shorter length than most other columns, and they are packed with smaller particles that are typically 3 μm in diameter. Advantages include increased sensitivity, decreased analysis time, decreased mobile phase usage and increased reproducibility.

Preparatory columns are utilized when the objective is to prepare bulk (milligrams) of sample for laboratory preparatory applications. A preparatory column usually has a large column diameter, which is designed to facilitate large volume injections into the HPLC system.

Accessories important to mention are the backpressure regulator and the fraction collector. The backpressure regulator is placed immediately posterior to the HPLC detector. It is designed to apply constant pressure to the detector outlet, which prevents the formation of air bubbles within the system. This, in turn, improves chromatographic baseline stability. It is usually devised to operate regardless of flow rate, mobile phase, or viscosity. The fraction collector is an automated device that collects uniform increments of the HPLC output.

6.4.1.7.5.3 Detectors

Detectors are the components that emit a response due to the eluting sample compound and subsequently signals a peak on the chromatogram. It is positioned immediately posterior to the stationary phase in order to detect the compounds as they elute from the column. The bandwidth and height of the peaks may usually be adjusted using the coarse and fine-tuning controls, and the detection and sensitivity parameters may also be controlled (in most cases). There are many types of detectors as given below;

6.4.1.7.5.3.1 Ultra-Violet (UV) detectors

Ultra - Violet detectors measure the ability of a sample to absorb light (Fig. 9). This can be accomplished at one or several wavelengths. Generally measures in the UV area

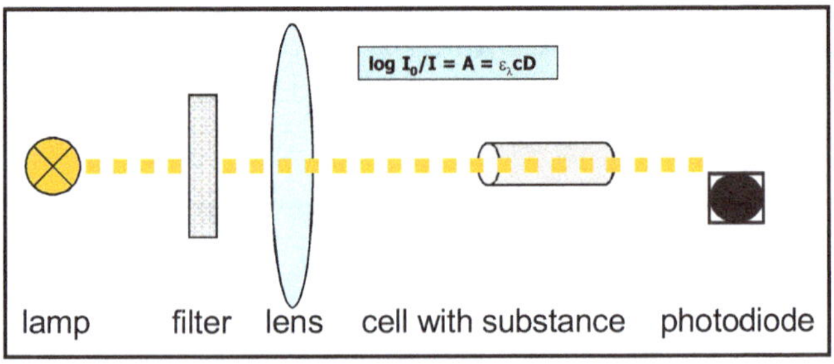

Fig. 9. UV/VIS Detector

(<350 nm, D_2 lamp), occasionally in the visible (VIS) area (>350 nm, Tungsten Halogen lamp). It is possible to calculate the concentration of the substance (after calibration) from the measured absorption.

A) *Fixed Wavelength:* measures at one wavelength, usually 254 nm.

B) *Variable Wavelength:* measures at one wavelength at a time but can detect over a wide range of wavelengths.

C) *Diode Array:* measures a spectrum of wavelengths simultaneously. UV detectors have sensitivity to approximately 10-8 or 10 -9 µg/ml.

D) *Spectro-photometer:* wavelength can be freely selected, wavelength changes while in operation, and spectra can be recorded in stop flow mode.

6.4.1.7.5.3.2 Refractive index (RI) detectors

Refractive index (RI) detectors measure the ability of sample molecules to bend or refract light. This property for each molecule or compound is called its refractive index. For most RI detectors, light proceeds through a bi-modular flow-cell to a photodetector. One channel of the flow-cell directs the mobile phase passing through the column while the other directs only the mobile phase. Detection occurs when the light is bent due to samples eluting from the column, and this is read as a disparity between the two channels.

6.4.1.7.5.3.3 Fluorescent detectors

Fluorescent detectors measure the ability of a compound to absorb then re-emit light at given wavelengths. Each compound has a characteristic fluorescence. The excitation source passes through the flow-cell to a photodetector while a monochromator measures the emission wavelengths. It has sensitivity limit of 10-9 to 10-11 µg/ml.

6.4.1.7.5.4 Qualitative and quantitative analysis using HPLC

6.4.1.7.5.4.1 Qualitative Analysis

Identification of a compound by HPLC is accomplished by researching the literature and by trial and error. A sample of a known compound must be utilized in order to assure identification of the unknown compound. Identification of compounds can be assured by combining two or more detection methods.

1. The main identification criterion for a component is its retention time.
2. Retention time is always the same for a particular compound in a given set of condition and is reproducible.
3. Photo diode array detector (PDA) can detect collating peak having UV-Visible absorbance but it has also got its limitation.
4. The latest LC-MS detector can be used for qualitative as well as quantitative analysis.

6.4.1.7.5.4.2 Quantitative analysis

Quantification of compounds by HPLC is the process of determining the unknown concentration of a compound in a known solution. It involves injecting a series of known concentrations of the standard compound solution onto the HPLC for detection. The chromatograph of these known concentrations will give a series of peaks that correlate to the concentration of the compound injected.

Using the area of a triangle equation (A=1/2b x h) to calculate the area under each peak, a set of data is generated to develop a calibration curve. Graphing peak area vs. the concentration of the sample solution does this. Most graphs can be generated using a computer software program such as Excel or Cricket graph (Fig.10). From this graphing software, a best-fit line can be derived, and the equation of that line can be determined. This equation of a line, y = mx + b, generated by the data, is the calibration curve equation.

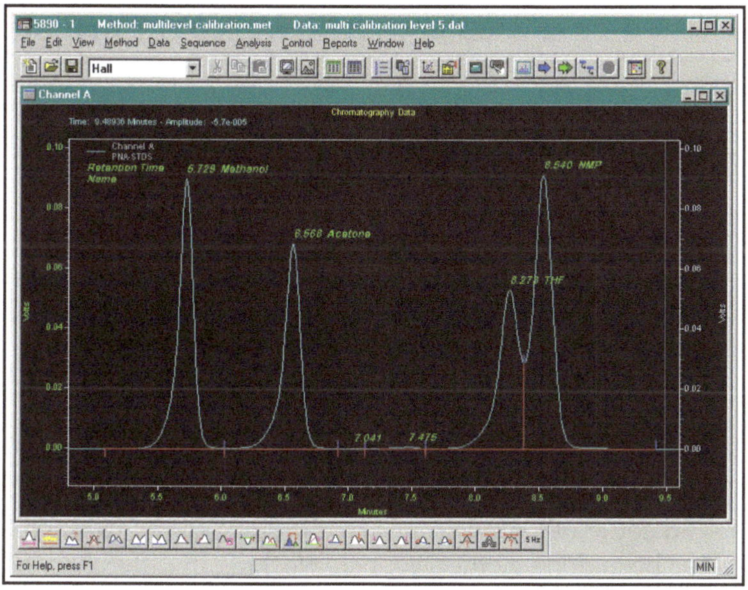

Fig. 10. Chromatogram

The equation of the line is then used in the following manner: A scientist injects a sample of unknown concentration x (x-axis of calibration curve) onto the HPLC; the chromatograph gives a peak output of area y (y-axis of the calibration curve). The area, y, is then in the equation of a line y=mx+b from the calibration curve, and the concentration is found by solving the equation for x.

This procedure, which is the most widely used, is known as the external standardization method (Fig. 11). Other quantitation methods, such as internal standardization, is the same that of above but only difference is that known amount of internal standard is added into standard and samples. In this the ratio of standard/ internal standard is plotted against the concentration.

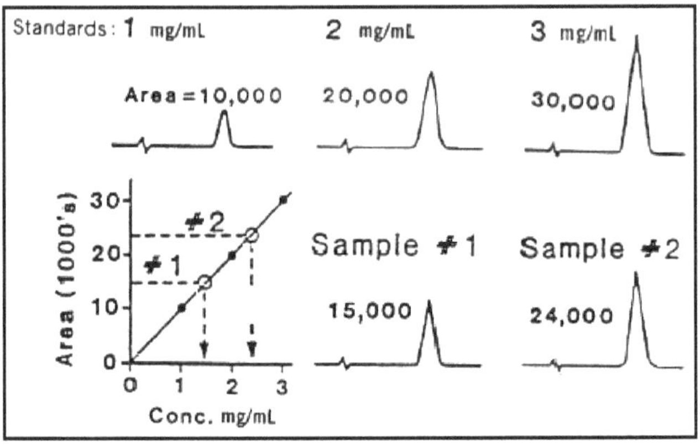

Fig 11. Calibration plot for external standard method

6.5 Extent of Residues of Insecticides in Different Commodities

The estimation of residues of insecticides in soils, crops and some other substrates by different analytical methods has been done by the scientists working in different residue laboratories of the country under supervised trials as well as the samples of produce collected from different markets. The results of some of the trials are listed in the table 12 to 15 (Gupta, 2001).

Table 12. Insecticide residues in soils

Insecticides ai/ha	Application	Initial deposit (ppm)	Residue ppm	
BHC 5 kg	Soil	3.527	480	0.442
Parathion 10 kg	Soil	4.566	90	0.008
Carbaryl 15 kg	Soil	88.5	90	BDL
Heptachlor 2.5 kg	Soil	2.89	90	0.08
Phorate 1.5 kg	Soil	1.81	45-60	0.19
Endosulfan 3.0 kg	Soil	2.51-2.89	45-60	0.54

BDL- Below detectable level

ppm - parts per million

Table 13. Insecticide residues in crops at harvest

Insecticide ai/ha	Application	Crop	Residue (ppm)
BHC 5kg	Soil	Maize	0.197
		Beetroot	0.279
		Radish	0.362
		Carrot	0.764
Parathion 10 kg	Soil	Maize	0.024
		Beetroot	0.086
		Radish	0.101
		Carrot	0.054
Carbaryl 15 kg	Soil	Maize	BDL
		Beetroot	BDL
		Radish	BDL
		Carrot	BDL
Heptachlor 2.5 kg	Soil	Sweet potato	BDL
		Onion	BDL
		Potato	BDL
Phorate 1.5 kg	Soil	Sweet potato	0.15
		Potato	0.18
		Onion	0.09
		Cauliflower curd	BDL

Table 14. Residues of some insecticides in different crops at harvest time

Insecticides	Dose ai/ha	Applic- ation	Crop	Residue (ppm)
Chlorpyriphos	0.1%	On crop	Ridge gourd fruits	2.47 (7 days)
Demeton –O-methyl	0.03%	-do-	Cabbage head	0.50 (7days)
		-do-	Brinjal fruits	BDL (15 days)
Diazinon	0.05%	-do-	Bhindi fruits	0.93 (7 days)
				0.05 (15 days)
		-do-	Maize grains	BDL

Contd...

Table 14. Contd...

Endosulfan	0.1%	-do-	Brinjal fruits	0.88 (7 days)
		-do-	Bhindi fruits	0.98 (7 days)
		-do-	Maize grains	BDL
		-do-	Ridge gourd fruits	0.71 (7 days)
Ethion	0.05%	-do-	Ridge gourd fruits	0.33 (7 days)
		-do-	Long melon fruits	0.13 (7 days)
Fenitrothion	0.07%	-do-	Cabbage	0.23 (7 days)
		-do-	Bhindi fruits	0.32 (7 days)
		-do-	Brinjal fruits	0.55 (7 days)
		-do-	Ridge gourd fruits	2.15 (7 days)
		-do-	Maize grains	BDL
Heptachlor	5 kg	-do-	Gram pods and grains	BDL
		-do-	Maize grains	BDL
Quinalphos	0.025% 0.5 kg	-do-	Brinjal fruits	2.09 (7 days)
		-do-	Maize grains	BDL
Thiometon	0.05%	-do-	Brinjal fruits	0.15 (7 days) BDL (15 days)
Tetradifon	0.05%	-do-	Long melon fruits	4.53 (7 days) 1.30 (15 days)
Toxaphene	1 kg	-do-	Maize grains	BDL
		-do-	Mustard seeds	BDL
	0.05%	-do-	Long melon fruits	2.40 (7 days) 0.13 (15 days)
BHC	0.5 kg	-do-	Cabbage head	0.05 (7 days)
		-do-	Gram pods and seeds	BDL
		-do-	Maize	BDL in grains
	0.05%	-do-	Brinjal fruits	1.66 (7 days)
Chlordane	0.5 kg	-do-	Gram	BDL

Contd...

Table 14. Contd...

		-do-	Maize	BDL
		-do-	Brinjal	0.77 (7 days)
Dicofol	0.05%	-do-	Ridge gourd fruits	1.08 (7 days)
		-do-	Long melon fruits	0.857 (7 days)
Dichlorvos	0.03%	-do-	Brinjal fruits	0.24 (7 days)
	0.05%	-do-	Ridge gourd fruits	2.93 (7 days)
	0.05%	-do-	Long melon fruits	0.087 (7 days)
Dimethoate	0.03%	-do-	Cabbage	0.84 (7 days)
	0.03%	-do-	Ridge gourd	0.017 (15 days)
Lindane	0.1%	-do-	Maize grains	BDL
		-do-	Brinjal fruits	2.39 (7 days) 0.76 (15 days)
		-do-	Bhindi fruits	2.85 (7 days) 0.68 (15 days)
Malathion	0.5 kg (dust)	-do-	Cabbage	BDL (7 days)
		-do-	Mustard seeds	BDL
		-do-	Brinjal fruits	0.73 (7 days) BDL (15 days)
		-do-	Bhindi fruits	3.23 (7 days) BDL (15 days)
Carbaryl	1.0 kg (dust)	-do-	Cabbage head	1.49 (7 days)
		-do-	Gram	1.57 in pod BDL in seeds
	0.1%	-do-	Cabbage head	3.51 (7 days)
		-do-	Maize grains	BDL
	0.15%	-do-	Ridge gourd	5.97 (7 days) 3.33 (15 days)
	0.1%	-do-	Long melon	4.47 (7 days) 0.30 (15 days)

Contd...

Table 14. Contd...

		-do-	Brinjal	6.2 (7 days) 0.9 (15 days)
		-do-	Barley grains	BDL
Malathion	0.5%	-do-	Ridge gourd fruits	BDL (7 days)
		-do-	Long melon fruits	2.9 (7 days)
Methyl parathion	0.5 kg (dust)	-do-	Cabbage head	3.06 (7 days) BDL (15 days)
		-do-	Mustard seed	BDL
Phenthoate	0.5 kg (dust)	-do-	Cabbage head	1.65 (7 days) 0.38 (15 days)
	0.5 kg (dust)	-do-	Bhindi fruits	2.15 (7 days) 0.93 (15 days)
	0.05%	-do-	Maize grains	BDL
Phosalone	0.5 kg	-do-	Cabbage head	2.13 (7days) BDL (15 days)
		-do-	Brinjal fruits	2.17 (7 days) BDL (15 days)
	0.035%	-do-	Cabbage head	1.41 (7 days) 0.01 (15 days)
		-do-	Bhindi	0.21 (7 days) BDL (15 days)
	1.0 kg	-do-	Mustard pods	0.33 (30 days)
		-do-	Mustard seeds	BDL
Quinalphos	0.5 kg (dust)	-do-	Cabbage head	3.26 (7 dyas) BDL (15 days)
		-do-	Maize grains	BDL (15 days)
	0.025%	-do-	Cabbage head	0.98 (7 days) 0.21 (15 days)
Carbaryl	0.1%	-do-	Onion leaves and bulbs	2.1 (3 days) BDL (4 days)
Endosulfan	0.1%	-do-	Ber	1.62 (5 days) 1.02 (9 days)

Contd...

Table 14. Contd...

	0.07%	-do-	Soybean leaves	0.54 (7 days)
		-do-	Seeds	BDL
	0.105%	-do-	Sugarcane stem	1.42 (6 days) 0.79 (9 days)
Monocrotophos	0.072%	-do-	Sugarcane stem	1.47 (6 days) 0.71 (9 days)
		-do-	Soybean leaves	0.54 (7 days)
		-do-	Soybean seeds	BDL
Fenvalerate	0.01%	-do-	Soybean leaves	0.37 (7 days)
		-do-	Soybean seeds	BDL
Cypermethrin	0.01%	-do-	Soybean leaves	0.38 (7 days)
		-do-	Soybean seeds	BDL
Carbaryl	0.125%	-do-	Mustard dry pods	2.385 in 7 days
		-do-	Mustard seeds	BDL
Phosphamidon	0.031%	-do-	Mustard dry pods	0.325 (7 days)
		-do-	Mustard seeds	BDL
Dimethoate	0.05%	-do-	Mustard dry pods	0.563 (7 days)
		-do-	Mustard seeds	BDL
Permethrin	0.005%	-do-	Gram pods	0.048 (14 days)
		-do-	Gram seeds	BDL
Cypermethion	0.005%	-do-	Gram pods	0.031 (14 days)
		-do-	Gram seeds	BDL

Table 15. Waiting periods of insecticides on crops

Sl. No.	Insecticide	Crops	Waiting Period (days)
1.	Endosulfan	Cow pea, cauliflower, cabbage	10
		Tomato	8
		Ber	5

Contd...

Table 15. Contd...

		Bhindi, Brinjal, pea, checkpea	7
		Pigeon pea	6
		Soybean	15
2.	Lindane	Bhindi	6
		Onion	9
3.	Malathion	Brinjal	1
		Soybean	2
		Torai, bhindi, long melon, cabbage	3
		Tomato	5
		Onion	6
4.	Quinalphos	Cabbage	10
		Cauliflower	1
		Mustard, acid lime	15
5.	Dichlorvos	Torai	7
6.	Dimethoate	Torai, rough surface fruits and vegetable	10
		Smooth surface fruits and vegetable	3
7.	Fenitrothion	Cow pea, pea	10
		Cabbage	7
		Pigeon pea	15
		Cauliflower	8
		Mustard	3
8.	Metasystox	Mustard	15
9.	Monocrotophos	Cow pea, pea, cauliflower	10
		French bean, bhindi	7
		Soybean, green gram, acid lime	15

Contd...

Table 15. Contd...

10.	Methyl parathion	Smooth surface fruits and vegetable	2-4
		Rough surface fruits and vegetables	5-8
11.	Phorate	Cabbage and cauliflower	10
		Soybean	65
12.	Phosalone	Pea	5
		Cabbage	10
		Green gram	16
13.	Phosphamidon	Pea	5
		Okra, acid lime	10
14.	Quinalphos	Cabbage	10
		Cauliflower	1
		Mustard, acid lime	15
15.	Phenthoate	Cabbage	15
		Bhindi	21
16.	Carbaryl	Okra	1
		Pea, cauliflower	5
		Brinjal, torai, tomato and cabbage	7
		Mustard	10
		Chick pea	15
17.	Carbofuran	Soybean	65
18.	Cypermethrin	Bhindi	4
		Cabbage	6
19.	Fenvalerate	Bhindi, cabbage	9
20.	Decamethrin	Bhindi	1
		Cabbage	2

References

Azad Thakur, N.S. and Rajasekhara Rao, K. (2006). Pesticide use in India and its environmental implications. In : Environment and natural resources : Ecological and economic prospective (Eds. O.P. Singh). Regency Publications, New Delhi. pp161-191.

Benson, W.R. and Finochiero, J.M. (1965). *J. Assoc. Off. Anal. Chem.*, 48:676.

Edelman (1962). *Assoc. Off. Agric. Chemi.*, 672.

Gupta, H.C.L. (2001). Persistence, residues and tolerances of insecticides. In: Insecticides: Toxicology and uses. Published by Agrotech Publishing Academy, Udaipur, India, pp 221-264.

Gunther, F.A. and Blinns, R.C. (1955). Interscience Publishers, New York.

Gunther, F.A. and Blinn, R.C. and Ott, D.E.(1966). *Bull. Environ. Contam. Toxicolo.*, 1:237

Jones, L.R. and Riddich, J. A. (1952). *Anal, Chem.*, 24: 569

Lykken, L. (1963). *Residues Rev.* 3:19.

Schafer, M.L., Busch, .K.A. and Compholl, J.E. (1963). *J. Dairy Sci.*, 46:1025.

7

Rapid Soil Testing Kit

Soil testing kit is developed to determine the available nutrient status of soil by rapid test methods. With this system of soil testing, a farmer can test his own soil as often and, in as many places as he feels necessary, and after gauging the need of his soil, he can apply the fertilizers as required. The present mode of soil testing in laboratories usually takes few days and sometime weeks and the farmers may have to travel long distances to submit the soil sample.

The soil testing kit available in the market in India is a box that contains thc following:

1. A booklet that describes the methods for determination of pH, organic carbon, available phosphorus, available potassium, ammonium and nitrate as well as interpretation of results. It also describes the procedure for collection of soil sample, liming, functions and deficiency symptoms of nutrient elements, etc.
2. Reagents.
3. Colour chart, scoops, syringes, beakers, and graduated test tubes with test tube stand, wash bottles, filter paper, droppers, spoon, and funnel. Glass marking pencil, glass rods, etc.

Another kind of kit named soil, plant, and water testing kit is also available. Its contents are:

1. A booklet (manual) that describes the methods for determination of pH, electrical conductance, available phosphorus, available potassium, nitrates. Also, it describes the method for determination of soil salinity, water quality tests and plant tissue tests.
2. Reagents.
3. Battery operated conductivity meter.
4. Other equipments

The soil testing kit can easily be transported to the field and soils may be tested there. In no soil testing kit the names of reagents are written. Instead, they are marked by numerical number. A typical soil test kit and its sample procedures is being described for easy understanding (Trakhtenburg, 2001).

7.1 Preparation and Extraction of Soil Samples

7.1.1. With distilled water the extraction tube from the Soil Test Kit is filled to the 30 ml line.

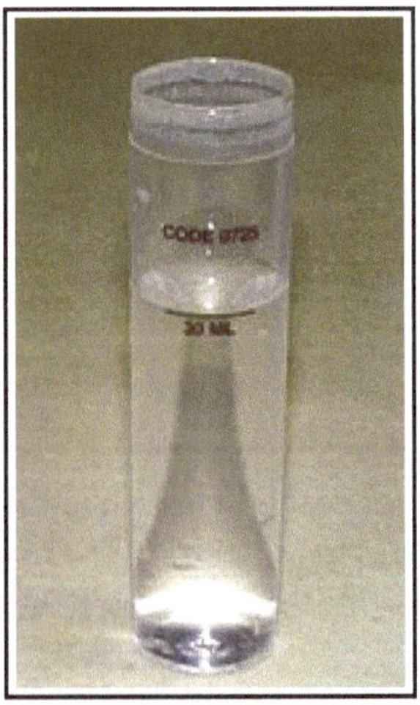

7.1.2. Add 2 Flock-Extractant tablets. Cap the tube and mix well until both tablets have disintegrated.

7.1.3. Remove the cap and add one heaping spoonful of dry, sieved soil.

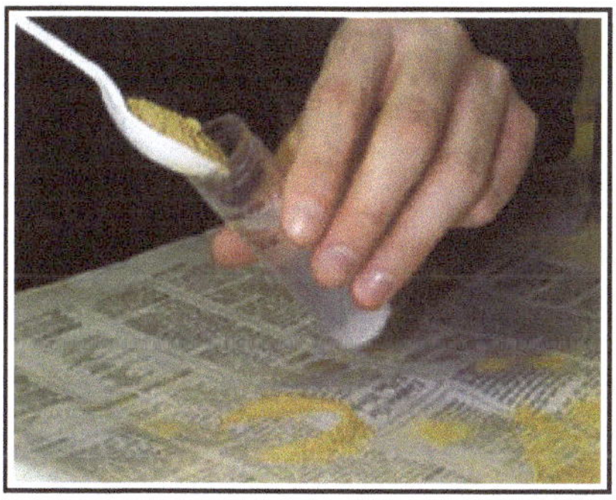

7.1.4. Cap the tube and shake for one minute.

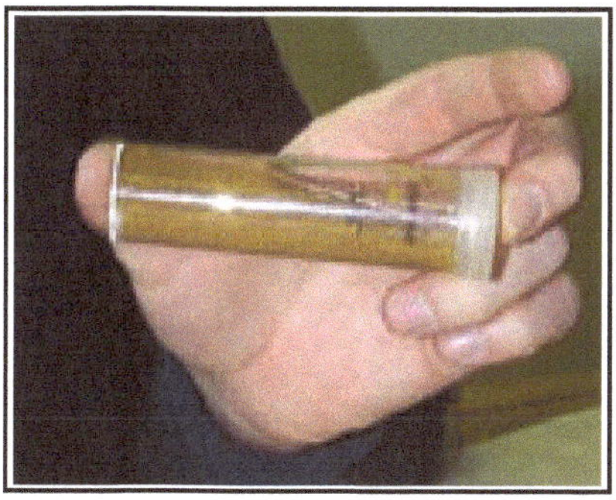

7.1.5. Let the tube stand until the soil settles out (usually about 5 minutes). The clear solution above the soil is used for the nitrate (N), phosphorus (P), and potassium (K) tests.

7.2 Test for Nitrate Nitrogen (N)

7.2.1 Use the pipette to transfer the clear solution above the soil to one of the test tubes in the Soil Test Kit until the tube is filled to the shoulder. (If more solution is needed, repeat Part 1).

7.2.2. Add one Nitrate Tablet. Be sure that all the pieces of the tablet are added to the test tube and try not to touch the tablet as you place it into the tube. Cap and mix until the tablet disintegrates.

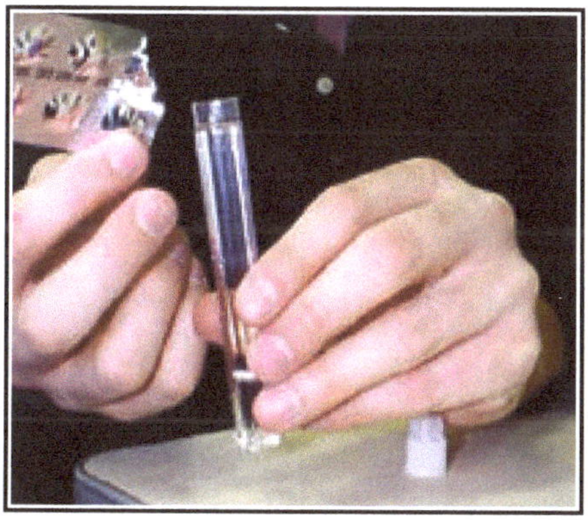

7.2.3. Rest the test tube in a cup or beaker. Wait 5 minutes for color to develop. (Do not wait longer than 10 minutes).

7.2.4. Compare the pink color of the solution to the Nitrogen Color Chart in the Soil Test Kit. Record your results (High, Medium, Low, or None) on the Soil Fertility Data Work Sheet.

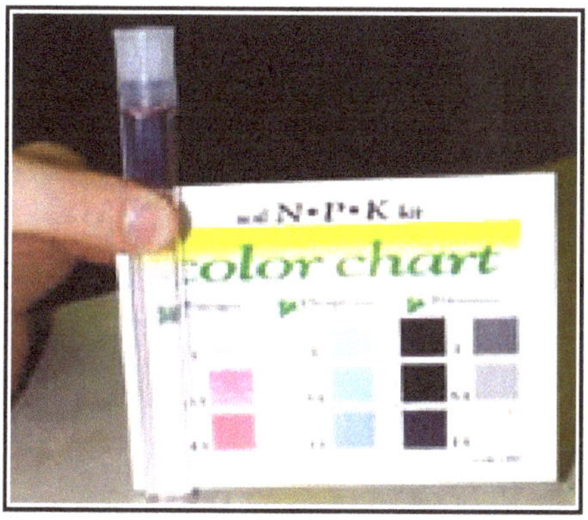

7.2.5. Discard the solution and wash the tube with distilled water.

7.2.6.Repeat this procedure with the liquid from each of the soil samples. Be sure to rinse the pipette and tube with distilled water after they are used.

7.3 Follow Same Procedures for Testing P & K.

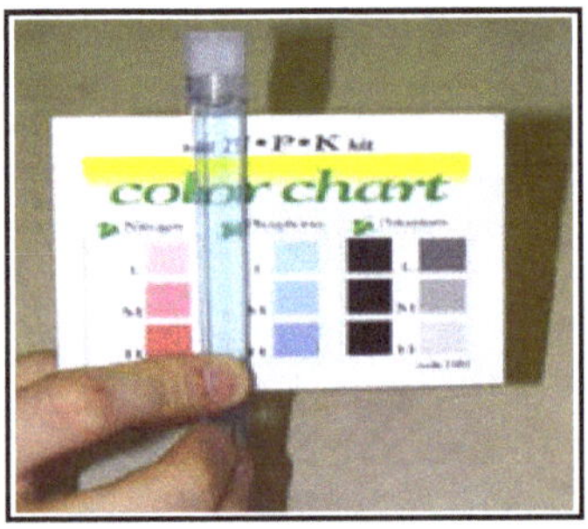

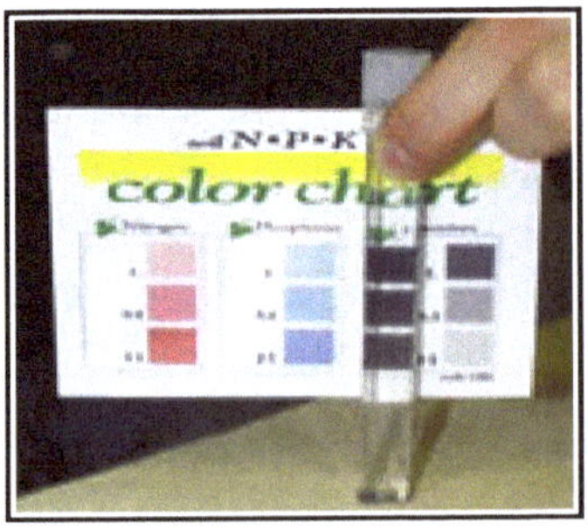

Reference

Trankhtenberg, I. (2001). In Soil Science Educational Home Series. (URL : izolda@ltpmail.gsfc.nasa.gov).

— • —

8

Establishing A Soil Testing Laboratory

To set up a soil-testing laboratory the minimum requirements are as follows:

8.1 Laboratory floor area

Minimum four rooms are needed. They are as follows:

1. Office room, floor size: 18'x 12'

This room is used for registration of the soil samples received, recording recommendations along with soil test results in laboratory book and in report sheet, their dispatch, accounts etc.

2. Soil sample preparation and storage room, floor size: 18' x 22'

This room is used for drying soil samples in shade, grinding, mixing, sieving, storage in box and placing box in storage racks.

3. Laboratory room, floor size: 18' x 25'

This room is used for weighing the soil samples, preparation of distilled water, chemical analysis, drying glasswares in a hot air oven and washing the glasswares, Exhaust fans are a must for this room to remove all kinds of fumes.

4. Instrument room, floor size: 18' x 9'

The instrument viz., pH meter, conductivity meter, photoelectric colorimeter, flame photometer and refrigerator are placed in this room and used. Thus, the instruments are saved from attack of acid fumes and dusts.

8.2 Equipments/ Instruments

1. pH meter
2. Conductivity meter
3. Photometric colorimeter
4. Flame photometer
5. Nitrogen distillation set
6. Distilled water plant
7. Mechanical shaker
8. Balance 0.1 mg sensitivity
9. Top laoding balance, 1 kg capacity
10. Wooden morter and pestle
11. Brass sieves (20 and 80 mesh)
12. Hot air oven
13. Refrigerator, 165 liter
14. Pocket calculator
15. Spade/Auger
16. Water vacuum suction pump
17. Soil testing kit

8.3 Glasswares

1. *Beakers:* 50, 100, 250, 500, and 1000 ml capacities.
2. *Conical flasks:* 150, 250, 500, and 1000 ml capacities.

3. *Volumetric flasks:* 25, 50, 100, 250, 500, and 1000 ml capacities.
4. *Measuring cylinders:* 10, 50, 100, 250, 500, and 1000 ml capacities.
5. *Graduated pipette:* 1, 5, and 10 ml capacities.
6. *Bulb type pipette:* 10, 25, and 50 ml capacities.
7. Glass rods
8. Funnels
9. Burette: 10, 25 and 50 ml capacities.
10. Reagent bottles
11. Watch glasses

8.4 Chemicals

1. Sulphuric acid
2. Hydrochloric acid
3. Orthophosphoric acid
4. Oxalic acid
5. Sodium hydroxide
6. Ammonium hydroxide
7. Potassium dichromate
8. Potassium permanganate
9. Potassium chromate
10. Potassium chloride
11. Ammonium chloride
12. Ammonium fluoride
13. Ammonium molybdate
14. Ammonium acetate
15. Activated charcoal (phosphorus free)
16. Buffer tablets (pH 4.0, 7.0, 9.2)

17. Butyl alcohol
18. Calcium acetate
19. Calcium chloride
20. Calcium sulphate
21. Calcium carbonate
22. Diphenylamine
23. 2,4 dinitrophenol
24. EDTA
25. Erichrom Black T
26. Ferrous ammonium sulphate
27. Hydroxyl amine hydrochloride
28. Methyl orange
29. Methyl red
30. Methanol
31. p-nitrophenol
32. Phenolphthalein
33. Sodium bicarbonate
34. Stannous chloride
35. Tri-ethanolamine

8.5 Others

1. Whatman No. 42 and 1 filter paper
2. Rubber corks to fit tightly the mouth of 250 ml conical flasks
3. Wash bottles
4. Polythene boxes, 500 ml capacity, to store soil samples
5. Burette stands
6. Pipette stands
7. Funnel holders

8. Beaker stands
9. Storage rack to place the box containing soil samples
10. Brush
11. Jerrycan to store distilled water
12. Bucket
13. Polythene sheet
14. Ball point pen or lead pencil
15. Hard paper tags
16. Rubber rings
17. Glass marking pen
18. Register book
19. Leaflets having the description of procedure of soil sample collection
20. Soil test report sheet
21. Voltage stabilizers
22. Exhaust fan
23. Graph paper
24. Scale (inches)
25. White writing papers

8.6 Staff

The minimum staff requirement may be as follows:

1.	Assistant Soil Chemist (Office-in -charge)	1
2.	Soil Analysts	2
3.	Laboratory Attendants	4
4.	Junior Assistant	1

Plates Contd...

Water distillation set

Grinding machine

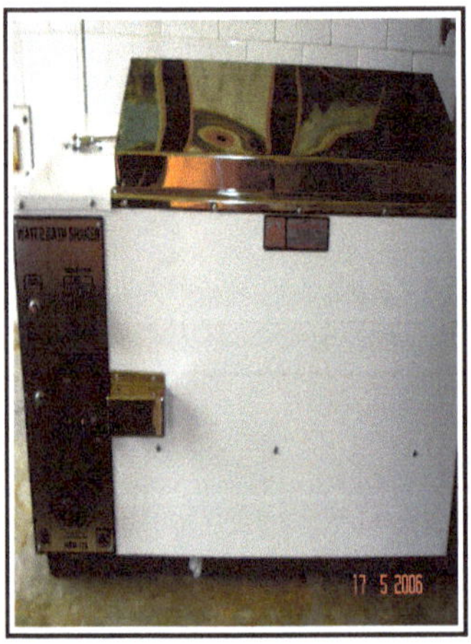

Water bath

Centrifuge

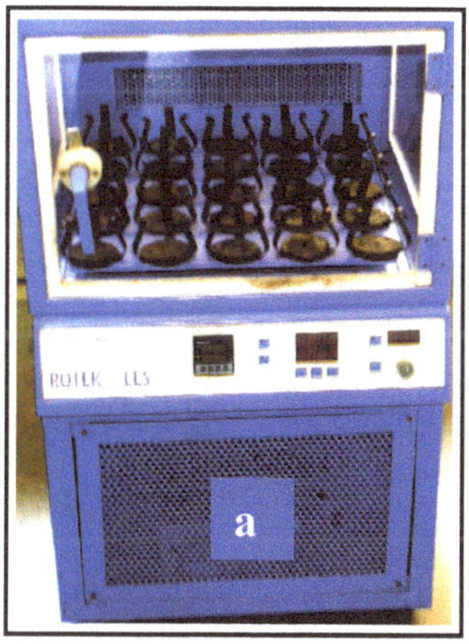

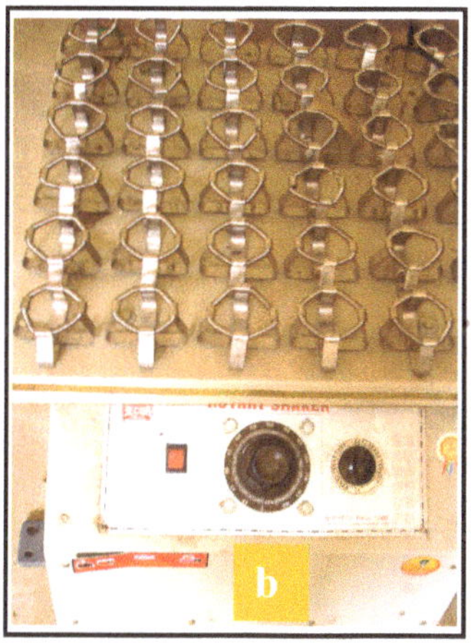

Shakers (a) Auto (b) Mechanical

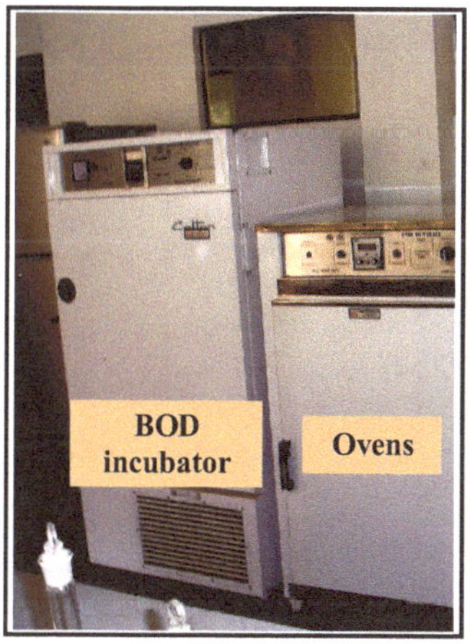

Ovens and BOD Incubators

pH meter

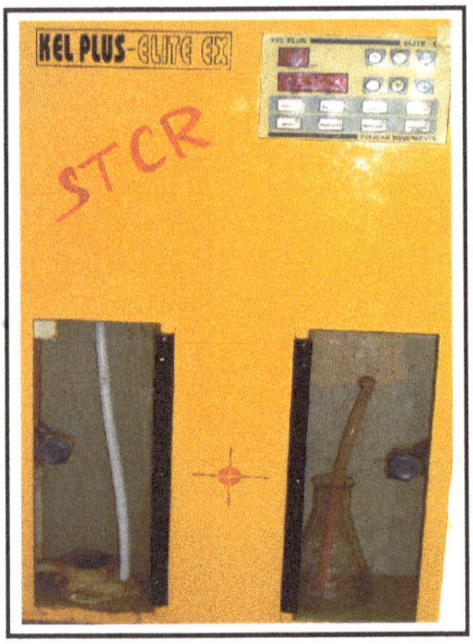

Distillation unit for N

Digestion unit

Computererised spectrophotometer

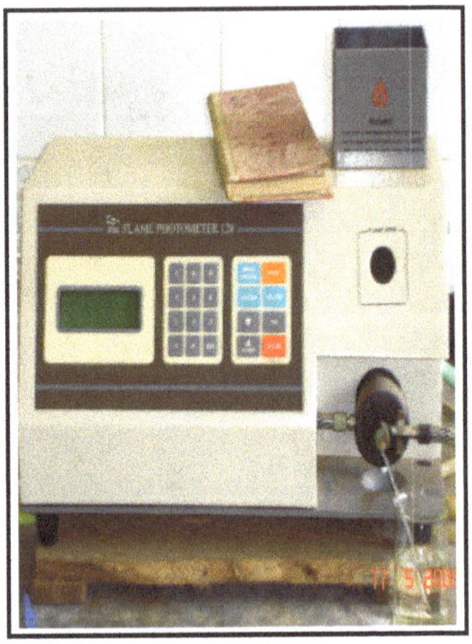

Flame photometer

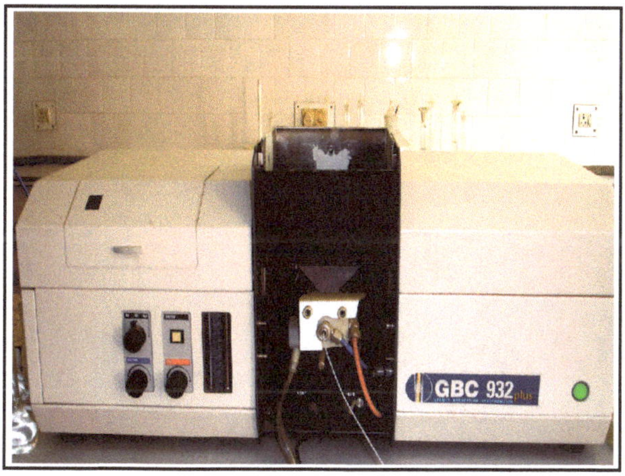

Atomic absorption spectrophotometer(AAS)

Filtration process in progress

Sieves

Pressure plate membrane apparatus

9

Instruments in a Soil Testing Laboratory

The analytical work in a soil-testing laboratory mainly involves standard chemical methods, suitably modified to permit handling of a large number of soil samples with the required degree of accuracy and speed. Many of the analytical operations are carried out more conveniently with help of common as well as sophisticated instruments.

The various instrumental measurements followed in soil and water-testing work in the laboratory can be broadly divided into two groups:

1) Determination of pH and electrical conductivity
2) Colorimetric, spectrophotometric

9.1 pH meter

There are two main kinds of pH meter, the potentiometric and the direct reading type. The direct reading type of instrument, although probably less accurate than the potentiometric is used almost exclusively in the soil laboratories all over the world. The typical pH meter consists of a glass membrane electrode and a reference electrode or a single probe combination electrode. The most commonly used reference electrode, which incorporates a salt bridge filled with saturated KCl solution. The electrode always works in a liquid environment; however there are dry pH electrodes that can be stored without liquid and those are often used for field works. The values of pH are determined from the sequential measurements of electrical potential (emf) of the cell. The pH cell consists of an electrode sensitive to hydrogen ions (platinum, glass, quinhydrone etc) and a reference electrode and can be broadly represented as:

H^+ sensitive electrode	Reference buffer or test solution	salt bridge	Reference electrode

In the conventional instruments the measuring electrode is of glass and the reference one is a calomel electrode.

9.1.1Glass Electrode

This consists of a very thin-walled bulb of H^+ responsive (soft) glass sealed to a stem of non-H^+ responsive glass. Both surfaces of the glass membrane are sensitive to changes in H^+ ion concentration. The inside of the bulb is filled with an electrolyte of high buffer capacity for example 0.1N solution of HCl. An internal reference electrode generally calomel one also dips into the test or buffer solution. When the glass bulb is dipped in a solution with H^+ ions concentration different from that of the solution within the electrode its outer inner surfaces attain a difference in electrical potential. This difference is a function of H^+ ion concentration of the outer solution, since the pH of the inner one is constant (unchanged).

9.1.2 Calomel Electrode

The electrode consists of mercury in contact with a solution of potassium chloride saturated with mercurous chloride (Hg_2Cl_2) within the glass bulb. It maintains a constant potential at a given temperature. In commercial models a paste of mercury and Hg_2Cl_2 contained in an inner tube connected to KCl solution in an outer jacket. The lead wire is connected to Hg_2Cl_2 paste through a column of mercury. The outer tubes in a fine capillary provide a salt bridge through the test solution to the glass electrode. pH meter with single (actually combined) electrode is also available as in case of digital types instruments.

9.1.3 Measurement of pH

A pH meter is an actually a direct current amplifier. That measures the emf, which appears across the electrodes upon being immersed in solution soil suspension or irrigation water. The glass and calomel electrodes are immersed in the test solution and the emf determined by the electron tube voltmeter. The meters graduated to read directly in pH unit alongwith the emf (milli-volt) scale. This method is based on the potential developed across the indicator or glass electrode by the difference in activity of H^+ ions in and out of the electrode, i.e., in the bathing solution. The potential, E, is defined by the Nernst equation:

$$E = (RT/nF) \log [K/(M)]$$

Where, R = Gas constant (8.313 Joules)

T = Absolute temperature (298°K or 25°C)

n = Valency of the metal ion (1 for H^+)

F = Faraday constant (= 96,500 coulombs)

(M) = Activity of the ion to be measured.

It is not possible to measure the half-cell potential, E, alone. So, another half-cell of a standard reference electrode or a calomel electrode must be used to measure the potential difference, in and out of the glass electrode, in conjunction with the reference electrode, when both are dipped in the solution under test.

Measurement of pH, on a pH meter, requires the initial standardization of the electrodes used. As such, both the electrodes have to be dipped in a solution of known pH [i.e., (H^+) = 1 g/L)], and the overall potential of the complete cell, E_o, equals the potential difference between the two electrodes, i.e., E-E_{cal} (where, E_{cal} stands for the standard potential of the reference electrode, i.e., 0.246 volt for the saturated electrode, containing saturated KCl solution).

$$E_0 = (RT/F) \log [K/1] - E_{cal}$$

When the electrodes are placed in a solution of known activity of H^+ ions, the potential difference, E_c, between the electrodes is given by:

$$E_c = (RT/F) \log [K/(H^+)] - E_{cal}$$

$$\text{Thus,} \quad E_c - E_0 = (RT/F) [\log \{K/(H^+\} - \log K] - E_{cal}$$

$$= - (RT/F) \log (H^+) = (RT/F).\, pH$$

$$= 0.0591 \text{ pH}$$

$$\text{Hence,} \quad pH = (E_c - E_0)/0.0591$$

The measurement is registered on a suitably calibrated Galvanometer or millimeter. The circuits of pH meter used in the soil testing laboratories are high impedance type, before taking the measurement, it is essential to set the instrument at the temperature of the test solution with the help of the temperature setting knob provided.

9.1.4 Care and Maintenance

The most delicate part of the pH meter is the glass electrode. It may crack or break if handled roughly or it may dry up when left out of water for a long period. In the later case the electrode should be immersed in 0.1N HCl and drained in the distilled water for 24 hrs or more and checked again. To identify the defective ones, the electrode jacks are short circuit with a special strap and terminal connections are provided with the instruments. The pH meter is switched on and sufficient time (10-15 min) is given for warming up. If the galvanometer needle reads the pH of 7.0 or can be adjusted to pH 7.0 with the standardized control in either pH range. The electronic side of the pH meter is in order and the defect is in the electrode and hence this may need replacement. It is equally essential to see that there is always some solid KCl in reference (calomel electrode). The film of the clay deposited on the electrode needs to be removed with the help of the swab of cotton and distilled water. Set the temperature control according to the buffer solution/ sample temperature or an automatic temperature compensates may be used. Most of the modern pH meter has a thermostatic temperature probe, which allows for automatic temperature correction, since pH varies somewhat with temperature. Dip the electrodes into the standard 4.0 pH buffer solution. Adjust the pointer or dial to show the theoretical pH values. Switch to the rest positive on and wash the electrodes. Repeat this process using a high pH (pH 7.00) buffer and then re-check with the low buffer. If the meter response shows a difference greater than 0.1-pH unit from the expected values indicates a faulty electrode. Glass electrode fail because of scratches, deterioration or accumulation of debris on the glass surface. Rejuvenate the electrode by alternating immersing in three times each in 0.1N HCl and 0.1 N NaOH. After rejuvenation soak in pH 7.0 buffers overnight rinse and store in pH 7.0 buffer.

9.2 Electrical Conductivity

The electrical conductance of a solution is measured with a Wheatstone bridge of which several portable models are available for field use.

9.2.1Principle

Ion being the carrier of electricity, electrical conductivity is a measure of the ability of a salt solution to carry electric current by the migration of ions under the influence of electric field. Electrical resistance is defined by the equation

$$E= IR$$

In which E is the electrical potential in volts, I is the current in amperes and R is the resistance in ohms. Electrical conductance, C, or conductivity of a solution in mhos is the reciprocal of resistance, R, in ohms,

$$C = I/R$$

For convenience, conductance is often expressed in millimhos (1000 times mhos). Specific conductance (L) of a solution is the conductance that would be measured at 25°C between electrode 1 cm^2 in cross section and place one cm apart, and may be visualized as the conductance across a cm^3, mhos/cm. Specific conductance may be measured with a cell of various dimensions by mean of cell constant. The cell constant, k, of a conductance cell is determined by measurement of the electrical conductance, C, of a standard KCl solution, and use the equation:

$$k = L/C$$

In which L is the known specific conductance of standard solution, usually 0.02 M KCl, mmhos/cm, and C is the conductance of the standard solution measured in given cell, mmhos.

The measured conductance, C, of a test solution in mmhos, multiplied by the cell constant gives the specific conductance, $L_{mmho/cm}$, of the test solution:

$$L = kC$$

9.2.2 Component of Conductivity Meter

The instrument used for this purpose is known as conductivity bridge a typical system consists of an alternating current (AC). Wheatstone bridge, a primary element of the conductivity cell and a null balance indicator (as in solubridge/ or an electric "eye" in the conductivity meter). The essential component of the cell is the two electrodes coated with platinum black, and rigidity set at a specific distance (5 mm or so) filled with inert electrolyte usually KCl or NaCl, agar is often used for jellification. With modern apparatus an automatic temperature compensation device is incorporated, but if not, then corrections have to be made to reduce the conductivity readings at 25^{0}C.

9.2.3 Care and Maintenance

The conductivity meter has a long life, and it rarely goes wrong unless ill handled. Non-linear response may be caused by a non-conditioned cell, a dirty cell, inadequate care in preparation of standards, inadequate attention to particular bubbles from the cell during standardization, a defective cell, inadequate precise control of temperature, or inadequate rinsing of the apparatus between standards. If it does, the metallic cover of the meter may be unscrewed and the internal wiring examine for any loose contact or the vacuum tube may be checked. Quite often, the trouble is found in the conductivity cell. Sometimes due to careless handling or inadequate washing, a film of clay is deposited on the electrodes. Repeated washings with distilled water or two can remove it or three washings with nitric acid followed by distilled water. If the cell is so dirty that bubbles always form at the top of the cell when a receptacle of water is placed on the apparatus, ethanol or 1 N NaOH may be used to attempt cleaning. Neither should be left on the cell for more than two or three minutes. Temperature need to be accurate within 0.001-0.005K to achieve the required reproducibility of data. When apparatus is not being used, the cells should be immersed in water.

9.3 Colorimetric Estimation

Analysis is based on the measurement of intensity of radiant energy after it passes through a coloured solution.

The procedure consists of passing a monochromatic light beam of known intensity through the sample solutions and determining the intensity of the transmitted beam with the help of a photoelectric cell. In other words, colorimetric can be considered as absorption spectrophotometric in the visible range of light.

The quantitative law basic to colorimetric analysis is Beer's law, which may be stated as follows:

$$\text{Log } I_o / It = k.C \text{....................1}$$

Where, Io = Intensity of incident radiation

It = Intensity of transmitted radiation.

C = Concentration in solution

k = A constant

The term 1 is the absorbance or optical density of the sample. It must be realized that the above relationship holds good only where absorption of the incident light occurs in the soltuion i.e when the light is of the wavelength at which the sample absorbs. Thus, monochromation (to a very narrow wave band) of the incident light through suitable filter or grating is an essential requisite of colorimetric/ absorption spectrophotometry.

9.3.1 Types of Colorimeter

Instruments of absorption photometry may be classified as visual comparators, filter photometers and spectrophotometers in terms of construction one also recognizes the differences between single and double beam light pass and whether the particular instrument is direct reading or implies balanced circuit. Special features include double monochromation and double wavelength monochromators plus arrangement for automatic recording. In the selection of an instrument, one should consider the initial cost, maintenance, flexibility of operation and adaptation to special situations. A typical colorimeter will have

1. A light source usually a tungsten lamp
2. A monochromatic (filter, prism or diffraction grating)
3. Absorption cell
4. Photometry system for measurement of intensity of the light beam in terms electric current generated.

9.3.2 Care and Maintenance

No attempt should ever be made to touch the galvanometer assembly of the photoelectric colorimeter. If accidentally damaged by mechanical shock or ingress of corrosive liquid, a competent person should examine the instrument. When the electric bulb (usually 100w) fails to light and if there is dangling wire the bulb has been damaged, beyond repair and needs replacement. If the bulb filament is intact, any loose connection to be set right. Care must be taken while replacing the projection bulb with one of the same wattage (100w) and of the same dimensions. If photocells fitted in the instrument loose their sensitivity for any reason, they may be surface cleaned or replaced as may be necessary but never to be opened. It is always advisable to use an automatic voltage stabilizer in the AC supply line. An important part of the instrument is the test tube (absorption cell) for holding the colour solution to be measured. When filled with the liquid the test tube acts as a cylindrical lens, converging the light to sharp line on a photoelectric cell. If it is not circular in cross section its focal length will change at different angles and so also will the intensity of the light falling on a photocell. A perfectly circular tube will not show any deflection of the galvanometer needle, at any angle of rotation when filled with distill water. It is also equally important that the walls of the tube are absolutely clear and free from scratches and no liquid is sticking to the outer side while placing it in the socket provided in the colorimeter.

9.4 Flame Photometry

When a substance is heated, it emits radiant energy. The emission becomes stronger with greater excitation of the molecules/atoms. This energy (electromagnetic radiation) is composed of radiation of different wavelengths; the assemblage of radiation is the emission spectrum of the substance. There are three kinds of emission spectra:

1) *Continuous spectrum:* given out by incandescent solids, consisting of continuous wavelength range, where individual lines are absent

2) *Band spectrum:* emitted by excited molecules, consisting of individuals bands which are actually composed of groups of lines very close to one another;

3) *Line spectrum:* originating from excited atoms or atomic ions (which exclude polyatomic ions or radicals). These spectra consist of distinct and often widely spaced lines.

Analytical flame photometry is based on the measurement of the intensity of the characteristics line emission of the element to be determined (Jackson, 1973). When a solution of a salt is sprayed into a flame (acetylene, propane or liquefied petroleum gas) the salt gets separated into its component atoms to higher energy levels. Actually the orbital electrons are shifted to higher planes from their normal orientation. When the electrons return back to the ground or unexcited state, they emit characteristic radiation. Since the excitation can be to different levels, the electrometric radiation of several wavelengths can be emitted. However, the intensity of the wavelength corresponding to the most probable transition will be the highest. For each element, such characteristic lines have already been very well identified. Each individual atom emits one quantum of radiation. Therefore, the intensity of radiation emitting from the flame (which must be otherwise steady and non-luminous) will be proportional to the number of atoms in the flame, that is, to the concentration of the particular element in the flame. This concentration is, in turn, directly related to the content of the element in the test solution.

The instrumental set-up for flame photometric analysis consists of three parts:

a) *Atomization assembly:* comprising a nebuliser (atomizer) to produce a fine mist from the solution and a burner which vaporizes the analytical (in atomized form);

b) *Monochromation system (filter, prism):* that separates out the analytical wavelength from the other radiations and

c) *Photometric system:* for measuring the intensity of the emitted radiation.

9.4.1 Various Types of Flame Photometer

A) *Single beam equipment* referred to as direct reading type comprises only one set of optics.

B) *Double beam equipment:* In this a second light path is provided for the radiation emitted by the internal standard element that is added to in a fixed amount to each test solution as well as calibration standard.

9.4.2 Care and Maintenance

Although a very sensitive instrument, flame photometers are usually fairly sturdy and give trouble free service for years if handled properly. The burner assembly has two external connections one leading in gas directly to the burner and other taking compressed air through a spraying nozzle (atomizer). The air stream carries with it to the burner flame a spray of the solution under test. If any choking or disconnection takes place it should be located and can be set right easily. The gas and air pressure must remain steady during the operations of the instrument and the connections and regulators should be checked before use. It is of utmost importance that only very clear solutions/ extracts are fed in to avoid any possible chocking of the capillary. Burners of flame photometers should be first turned off and then the compressed airline. When large salt concentrations or strong acid solutions are to be fed in (which should be avoided as far as possible) distilled water should be run for some time and the burner cleaned immediately after use. The galvanometer assembly must not be touched and any repairs should be carried out at a competent workshop.

9.5 Atomic Absorption Spectrophotometry

In atomic absorption spectrophotometry a sample, is aspirated and automised in a flame at higher energy states to excite the atoms and they then emit radiation by dropping down to the ground state. A light beam from a hollow cathode lamp or an electrode discharge lamp is directed through the flame into a monochromator and onto a detector that measures the amount of absorbed light. Atoms of elements in a sample, when excited emit light beam of the characteristic wavelength only the metal being determined. The intensity of light energy by the flame is directly proportional to the concentration of that metal in the sample.

9.5.1 Principle

Atoms are excited by thermal means to higher energy states and then they emit radiation by dropping down to less energetic states or to the ground state. The process can be summarized as:

$$M + hv \rightarrow M^* \rightarrow M + hv$$

Absorption Emission

Where M is the neutral atom, M* is the excited atom, h is Planck's constant and v is the frequency. The difference in energy between the two states E*-E is equal to

$\Delta E = hv = hc/\lambda$, where c is the speed of light and λ is the wavelength.

The number of atoms in a flame in a particular excited energy state (n *) and the number of atoms in the ground or unexcited state (n) are related by the following familiar boltzman relationship: $n^*/n = A \exp(-DE/KT)$

n^*/n values for most elements in flames of about 3,000°K are in the 10-10 to 10-4 range.

In absorption spectroscopy, the energy imparted to the atom (hv) is derived from the light source, whereas in emission spectroscopy this energy is primarily imparted to the atom in the atomization-excitation source. To get an absorption signal as strong as possible, AAS utilizes the resonance lines of those atoms in the ground (or near ground) state.

9.5.2 AAS - Components

- A light source that emits resonance line radiation,
- A sampling cell, usually a flame, into which the sample is fed as an aerosol,
- A monochromator that is used to isolate the absorbing resonance line from other nonabsorbent lines,
- A detector that measures the amount of absorption,
- Electronic amplification of the absorption signal, and
- A readout system that normally is a strip chart recorder, a digital display, a meter, or a printer.

9.5.3 Principle of Operation

- The sample aerosol is aspirated.
- Mixed with the gases before combustion.

- During combustion, the element is reduced to the atomic state and then can absorb radiation.
- Absorption of radiation is directly proportional to number of atoms present in ground state.
- Air-C_2H_2 flame (2,100-2,400°C) → used for elements with low ionization potentials that don't form refractory compounds.
- Nitrous oxide-acetylene is a hotter flame (2,600-2,800°C) → used for Al, As, Cr, Si, and Ti.

9.5.4 Types

- Single-beam
- Double- beam

These instruments vary in range from the most versatile and sophisticated. Choosing the right AAS unit will depend on one's workload, the variety of samples, and the funds available. The multielement sequential AAS unit is essentially a double beam system with upto a six cathode lamps mounted in a turret that can be programmed to determine the chosen elements sequentially. The dual channel double beam instrument allows for simultaneous determination of concentrations and ratios of two elements using identical atomization and flame parameters. A double beam system is preferred over a single beam system since it compensates for changes in lamp intensity or variations in electronic noise or detector sensitivity. Any change in the sample beam, other than absorption in the flame is cancelled out electronically by an identical change in signal from the reference beam. Since a double beam instrument sends only half of the source light intensity through the sample chamber, the eight-beam instrument, which utilizes all the light, will give better detection limits especially if the electronics are stable.

9.5.5 Nonflame Atomic Absorption Spectrophotometer

It is commonly known as AAS with graphite furnace atomizer (HGA) and very useful when (a) sample size is very restricted, and (b) lowest detection limit is needed. Typical sample volume is 20 to 100µL. If one is dealing with solid samples, 1 to 5 mg. When one is dealing with solutions, there is usually a three-stage procedures followed after pipetting the sample in HGA.

- Pipetting the sample into the HGA.
- Sample is dried at 110°C to remove the water.
- The temperature is raised to some value at which no atomization of the element in question occurs but where as much as possible of organics or salts is volatized.
- The temperature is raised sufficiently to atomize the element of interest. Typically this is between 2,100 and 2,900°C.

The entire procedure usually takes < 1 min per sample. The vastly lower detection limits for the HGA are due almost entirely to (a) increased residence time of the atoms in the optical path of the spectrometer, and (b) increased efficiency of atomization. However, spectral interferences and chemical interference are more common in this instrument.

9.5.6 Important Points Regarding Operation and Maintenance of AAS

Being one of the quite sophisticated, high precision and fairly expensive type of measuring instrument, the AAS, requires very careful handling and maintenance.

1. The light source is a critical component. There should not be any erratic fluctuation in intensity. Any fluctuation in the line voltage must be taken care of.

Lamps are generally operated at low current to improve linearity of response and maintain narrow emission lines, as much as possible.

2. While extinguishing the flame, the fuel supply should be first turned off before stopping the oxygen or airline. The oxygen (or air) and fuel pressure must be adjusted to the value recommended in the instruction manual furnished by the manufacturers. The width of the slit is usually set at 0.02-0.05mm as instructed in the working manual. The wavelength should also be set properly for each determination by making use of the relevant component like filter photometer monochromators.
3. The various aspects to be considered for optimum working condition are: the emission intensity, steady rate of atomization, the resolution of the spectral lines (or bands), from each other and from adjacent spectral features of other sample constituents, the magnitude of the background radiation, and steadiness of readings.
4. The AAS can give reliable and trouble free service only with careful handling. Erratic readings may come from fluctuation in gas and air pressure and electric supply or the lamps becoming weak due to prolonged use. If something goes wrong, a qualified mechanic or technician who should preferably be accredited by the manufacturers or their authorised agents for carrying out the repairing job should never open except the instrument.

9.6 Inductively Coupled Plasma Optical Emission Spectrophotommetry (ICP-OEC)

ICP-OES gained a strong foothold in environmental laboratories in the 1980s and 1990s due to several features over AAS such as

- Multielement analysis capability,
- Large dynamic linear range,
- Reduction of matrix interferences,
- Improved detection limits for refractory elements, and
- Enhancement of productivity.

9.6.1 ICP-OES Principle

- Sample constituents are vaporized→atomized→ excitation of atoms or ions.
- The excited atoms or ions will emit their specific line spectra.
- The intensities of these lines depend on the concentrations of their respective elements in the sample.
- Suitable detector analyzes emitted spectra.
- Plasma is used for vaporization, atomization, ionization and excitation of elements.
- Electrons and ions passing through the oscillating electromagnetic field flow at high acceleration rates inside the quartz tube space.
- Collisions between accelerated electrons (and ions) and ensuing unionized Ar gas cause further ionization.
- The collisions cause ohmic heating and give temperatures ranging from 6,000 to 10,000°K.

9.6.2 ICP-OES Major Components

- Sample introduction system
- Nebulizer
- Plasma generation assembly
- RF generator
- Detector

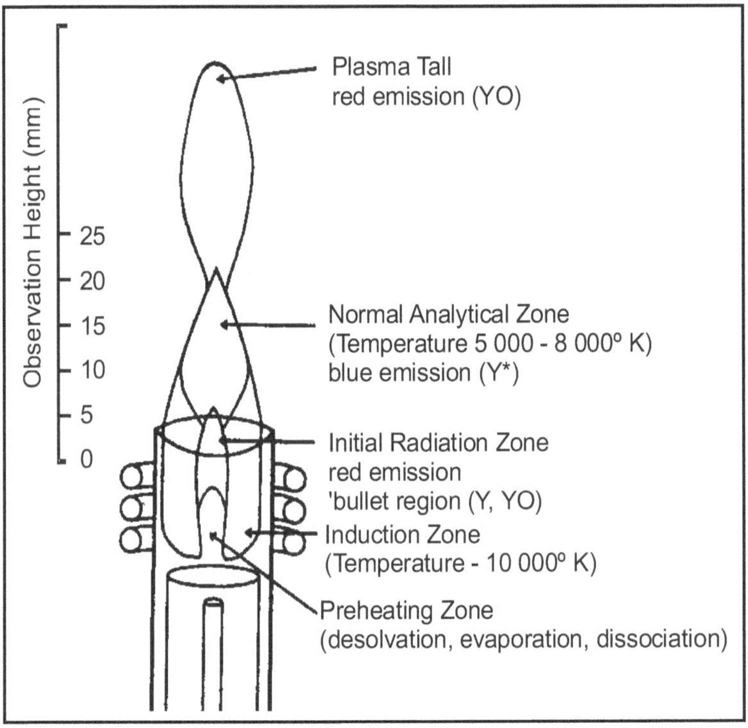

ICP-OES plasma torch

9.6.3 Classification of ICP-OES

- Method of measurement
  - Simultaneous
  - Sequential
- Viewing of plasma
  - Axial view
  - Radial view
- Detector
  - Photomultiplier tube
  - Solid state detector

Table.16 Comparison of performance in AAS vs. ICP-OES

S.No.	Parameters	AAS	ICP-OES
1	Detection limits	Furnace AAS detection limits are generally better	Detection limits for refractory elements (e.g. B, Ti, V, Al) are better by ICP than by flame AAS
		Detection limits for Group I elements (e.g. Na, K) are generally better by flame AAS than by ICP	Non metals such as phosphorus, sulphur, nitrogen, carbon, and the halogens (e.g. I, Cl, Br) can only be determined by ICP
2	Sample throughput	-	Sequential ICP becomes faster than AAS is approximately 6 elements/ sample for routine analysis
3	Unattended operation	Flame AAS cannot be left completely unattended for safety reasons	An ICP-OES instrument or graphite furnace AAS can be left to run overnight as no combustible gases are involved
4	Linear dynamic range	-	ICP has large linear dynamic range of about 105
		In AAS: Cu can be measured from 0.02 ppm to over 20 ppm	In ICP: Cu can be measured from 0.002 ppm to over 200 ppm
5	Interferences	In AAS, chemical interferences are high	ICP-OES is almost free from chemical interferences due to the high temperatures reached in a plasma
6	Spectral interferences	Minimum in AAS	In ICP-OES, greater possibility of spectral interference

Contd...

Table 16. Contd...

			In sequential ICP, can be overcome by selecting a different wavelength with similar detection limits
7	Precision		Recalibration is required in case of long duration analysis in case of AAS In ICP, long-term precision is quite high

— • —

10

Conversion Factors

Concentrations

1 microgramme = μg = 1 X 10^{-6} g

$\mu g\ cm^{-3}$ ($\mu g/ml$) = ppm (in solution)

ppb = parts per billion = $\mu g/L$

kg per $10^4\ m^2$ (= kg/ha) = 1.21 X lb/acre (thus lb/acre = 0.11 kg per $10^3\ m^2$)

g dm^{-3} = g/litre

cm^3 = cc = ml

dm^3 = litre (l)

centimole per kilogram (cmol/kg) = milliequivalents per 100 grams (me/100 g)

milligram per kilogram (mg/ kg or mg/ L) = parts per million (ppm)

% X 10000 = ppm

gram per 100 gram = percent, %

me/100 = me/100 X 10 X equ. wt. = ppm

Electrical Conductivity (EC)

1 siemens = 1S = 1/ohm = 1 mho

Siemens per meter (S/m) = millimhos per centimeter (mmhos/cm)

ppm = 0.64 X EC X 10^6 (for soil saturation extract)

Chemical Units

mole (M) = atomic, molecular, or ionic weight in gram

equivalent weight = 1 mole/ valence = 6 X 10^{23} reactive charges

molarity (1 M) = 1 mole per litre (1 M/L)

normality (1 N) = 1 equivalent weight per litre (1 eq wt/L)

milliequivalent = 1 me = 0.001 eq of l ml of N solution

Distance, Length & Thickness

mile = 1 mi = 5280 ft = 1609 m = 1.609 km

foot = 1 ft = 12 in. = 30.48 cm = 0.3048 m

inch = 1 in. = 2.54 cm = 25.4 mm

kilometer = 1 km = 1000 m = 3281 ft = 0.621 mi

meter = 1 m = 10 cm = 1.094 yd = 3.281 ft = 39.37 in.

centimeter = 1 cm = 0.01 m = 10 mm = 0.3937 in.

millimeter = 1 mm = 0.001 m = 0.1 cm = 0.03937 in.

micron = 1 μ = 10^{-6} m = 0.001 mm

Angstrom = 1 Å = 10^{-10} m = 10^{-7} mm

nanometer = 1 nm = 10^{-9} m

Area

square mile = 1 mi^2 = 640 ac = 259 ha

acre = 1 ac = 43560 ft^2 = 0.4047 ha

square kilometer = 1 km^2 = 100 ha = 247.1 ac = $(10^3\ m)^2$

hectare = 1 ha = 10,000 m^2 = 2.471 ac

square meter = 1 m^2 = 10.76 ft^2

Volume

gallon (liquid) = 1 gal = 4 liquid quarts = 231 $in.^3$ = 3.785 liters

cubic meter = 1 m^3 = 1000 litre = 28.38 bu = 35.31 ft^3

litre = 1 L = 1000 cm^3 = 1.057 liquid quarts = 61.02 $in.^3$

cubic centimeter = 1 cm^3 = 1 ml = 0.001 L = 0.0610 $in.^3$

cubic meter = 1 m^3 = acre-inch X 9.73 X 10^{-3}

cubic meter = 1 m^3 = cubic foot, ft^3 X 35.3

cubic meter = 1 m^3 = cubic inch, $in.^3$ X 6.10 X 10^4

Mass / Weight

ton (short) = 1 ton = 2000lb = 907.2 kg = 0.9072 metric tons

ton (metric) = 1 ton = 1000 kg = 1.102 short ton = 2205 lb

pound = 1 lb = 453.6 g

quintal = 1 q = 100 kg = 220.5 lb

kilogram = 1 kg = 1000 g = 2.205 lb

gram = 1 g = weight of 1 cm^3 of H_2O at 4°C

milligram = 1 mg = 0.001 g

Yield and Rate

kilograms/hectare = 1 kg/ha = 0.1 g/m^2 = 0.892 lb/ac

tons/acre = 1 ton/ac = 2000 lb/ac = 2.2417 metric tons/ha = 22.417 q/ha

pounds/acre = 1 lb/ac = 1.121 kg/ha = 1121 g/ha = 0.1121 g/m^2

bushels/acre = 1 bu/ac (wheat) = 60 lb/ac = 0.67 q/ha
= 1 bu/ac (corn or sorghum) = 56 lb/ac = 0.63 q/ha
= 1 bu/ac (oats) = 32 lb/ac = 0.36 q/ha

Temperature

Fahrenheit = $^{o}F = 32 + 1.8 \times {}^{o}C$

Celsius = $^{o}C = {}^{o}K - 273.15 = ({}^{o}F - 32)/1.8$

Plant Nutrient Conversion

$P \times 2.29 = P_2O_5$

$P_2O_5 \times 0.437 = P$

$K \times 1.2 = K_2O$

$K_2O \times 0.830 = K$

$Ca \times 1.39 = CaO$

$CaO \times 0.715 = Ca$

$Mg \times 1.60 = MgO$

$MgO \times 0.602 = Mg$

Sieve Mesh Size

Data for sieves showing comparison of US standard and I.M.M. sieves

Openings in mm	US Mesh No.	I.M.M Openings in mm	I.M.M Mesh No.
2.00	10	2.54	5
1.00	18	1.06	12
0.42	40	0.42	30
0.25	60	0.25	50
0.18	80	0.21	60
0.15	100	0.16	80
0.11	140	0.10	120
0.07	200	0.08	150
0.05	300	0.06	200

Notes:

1. The US standard and the Tyler sieves can be considered as being interchangeable.
2. The openings in mm are calculated as squares.
3. Sometimes sieve sizes are given in microns; for example, a 2000-micron sieve (or 2 mm) and 150 micron (or 0.15 mm), and so on.

11

Basic Analytical Calculations or Test Exercise

1. Calculate the number of g atoms of an element present in a given sample of Na weighing 25 g.

   Solution:

   We know the atomic weight of Na is 23

   Since, 1g atom of Na means 23 g of Na

   So, No. of g atom = Wt. of the element in g / Atomic wt. of the element

   = 25 g /23

   = 1.08 g atom of Na

2. Calculate the number of moles of NaOH in 400 g of NaOH.

   Solution :

   The number of moles in a given system is calculated as

   No. of moles = Wt. of the substance in g / Mol. Wt. of the substance

   = 400 g/40

   = 10 moles of NaOH.

3. Calculate the molarity of a solution of NaOH if 4 g of NaOH is to be dissolved in 1 lit. of solution.

Solution:

Molarity of a solution = No. of g mole / Volume of solution in ltr.

No. of moles = Wt. of the substance / Molecular wt.

= 4 / 40

= 0.1 mole

Molarity of NaOH = 0.1 / 1

= 0.1M

4. Calculate the formality of a solution in the above example.

Solution:

Since, the formula weight of NaOH is 40,

So, if 40 g of NaOH is present in 1 lit. then it is called a formal solution.

Formality of a solution = No. of g formula weight / Volume of solution in lit.

5. Calculate of Hydrogen ion concentration of a solution if the solution is 5.0×10^{-7} g ion/L

Solution:

Since pH = $-\log [H^+] = -\log [5.0 \times 10^{-7}]$

pH = $-[\log 5.0 + \log {}_{10}{}^{-7}]$

pH = $-[0.699+(-7.00)]$

pH = $-[0.699- 7.00]$

pH = 6.30

Since log axb = log a + log b

$\log 10^{-7}$ = -7.00

log 5.0 = 0.699 (From log table)

6. Calculate the pH of 0.01 N HCl

Solution :

In 0.01 N HCl = $[H^+] = 10^{-2}$ g ion / L

pH = $-\log_{10}^{-2} = 2$

Test yourself in the exercises given below:

Acidity

Q.1. What is the pH of a solution having $a_{H^+} = 4.2 \times 10^{-6}$ g H^+ ion/L

(Ans: 5.38)

Q.2. What is the a_{H^+} of a solution having pH=4.38

(Ans: 4.2×10^{-5} g ion/L)

Q.3. Calculate the pH of 0.05 M solution of sodium acetate (the dissociation content of acetic acid is 1.8×10^{-5})

(Ans: 8.72)

Q.4. Calculate the pH of 0.01N HCl and 0.01N CH_3COOH. The percentage dissociation is 12.5%

(Ans: 2)

Q.5. The pH of a solution is 10.7, calculate the H^+ ion concentration in the solution?

(Ans: 2.0×10^{-11})

Q.6. The H^+ ion concentration of solution is 5.0×10^{-7} g ion /L, calculate its pH and pOH

(Ans: pH=6.30, pOH = 7.69)

Q.7. In a solution, the H^+ ion concentration is 0.01 g ion /L, calculate its pH

(Ans: pH =2)

Q.8. In an aqueous solution , $(H^+) = 10^{-3}$ g ion /L, calculate the pH and pOH value of the solution

(Ans: pH=3, pOH =11)

Soil textural classes, bulk density, particle density and porosity

Q.9. Name the textural class of a soil having 55% clay, 32% silt and 13% sand.

(Ans: Clay)

Q.10. A sample of soil was sieved and had the size separates in material smaller than 2 mm determined by particle size (mechanical) analysis with the result

a) Coarse fragments screened out	53 g
b) Sand content	140 g
c) Silt content	38 g
d) Clay content	22 g
e) Total dry weight	253 g

Determine the soil textural class

(Ans: Sandy loam)

Q.11. Calculate the porosity of a soil having bulk density of 1.5 g/cc and particle density of 2.5 g/cc

(Ans: 40%)

Q.12. Calculate the particle density of a soil if the porosity is 40% and its bulk density is 1.5 g/cc

(Ans: 2.5 g/cc)

Q.13. On mechanical analysis soil was found to have the following composition

Sand	68.7%
Silt	10.0%
Clay	21.3%

What is the texture of the soil?

(Ans: Sandy clay loam)

Liming

Q.14. Calculate amount of limestone having 95% purity if 1000 kg pure $CaCO_3$ has been recommended?

(Ans:1053 kg)

Q.15. For liming an acid soil 1 t of pure $CaCO_3$ has been recommended. Find out the quantity of limestone with C.C.E. of 98%.

(Ans: 1.0204 t of limestone)

Q.16. For liming an acid soil , 5 t of pure $CaCO_3$ is needed. How much purer CaO will be needed?

(Ans: 2.79 t pure CaO)

Q.17. A limestone contains 40% CaO. Find the calcium carbonate equivalence of the liming material?

(Ans:71.6%)

Q.18. A limestone contains 20% MgO. Find the calcium carbonate equivalence of the lime stone.

(Ans:50%)

Q.19. A soil test report showed that the pH of a soil measure by SMP method is 5.0. Find out the quantity of a limestone with CCE of 90% for application in one hactare of a land.

(Ans:30.7tons)

Basic analysis

Q.20. How much gram of NaOH is required to make 0.1M solution of NaOH ?

(Ans: 4 g NaOH)

Q.21. A sample contains 0.2 g chloride gave 0.125 g ppt of AgCl. Calculate the percent chloride in the sample?

(Ans: 15.45%)

Q.22. In an analysis, the weight of ppt of Fe_2O_3 was found to be 0.315 g. Calculate the weight of Fe in the sample solution?

(Ans: 0.2209 g)

12

Glossary

Accelerated erosion: Erosion resulting from human involvement is called accelerated erosion because it is typically many times as fast as geological erosion.

Acid: Substance containing active hydrogen, which dissociates on solubilizing in water to produce H^+ ions. Lowry-Bronsted defined an acid is a substance that gives up a proton during chemical reaction. According to Lewis, an acid is a substance that can accept an electron pair and the base is a substance that can donate an electron pair.

The number of hydrogen ions that can be replaced by a metal determines the *basicity* of an acid. For example H_2SO_4 is a dibasic acid, HCl is a monobasic and H_3PO_4 is a tribasic acid.

Acid soil: Soil with a pH value less than 7.0.

Acidification: Process whereby soil becomes acid ($pH < 7$) because acid parent material is present or in regions with high rainfall, where soil leaching occurs. Acidification can be accelerated by human activities (use of fertilizers, deposition of industrial and vehicular pollutants).

Adsorbed cation: The cations held by the soil micelles are often called adsorbed cations. They are held in place by the attraction between the positive charge of the cation and the negative charge of the micelle.

Aggregate: Soil aggregate consisting of two or more soil particles bound together by various forces.

Aggregate stability: The ability of soil aggregates to rearrangement by various forces that may impact them, especially the effects of water, is called aggregate stability. The stability of aggregate is related to the soil texture, the kind of clay, the kinds of ions associated with the clay, the kind and amount of organic matter present, and the nature of the microbial population.

Aggregation: Process whereby primary soil particles (sand, silt, clay) are bound together, usually by natural forces and substances derived from *root exudates* and microbial activity. Soil aggregates are arranged to form soil peds, units of soil structure, classified by size (very fine, fine, medium, coarse and very coarse), shape (platy, prismatic, columnar, angular and subangular blocky, granular and crumb) and grade (single-grain, massive, weak, moderate, strong). From an agronomical point of view, the most important soil aggregates are in range 3 - 1 mm.

A horizon: The upper layer of soil is usually high in organic matter and darker in colour than the layers below. This upper layer is called the A horizon, or topsoil. The A horizon is the portion of soil most exposed to the weathering action of the sun, rain, wind, and ice and to the action of living things. The more easily decomposed materials weather away first, thus leaving the more resistant minerals concentrated in the topsoils.

Alluvium: Materials deposited by streams is called alluvium. The streams may be large or small, swift or slow. The bulk of alluvial material is transported during flood stage when heavy rains cause water to come pouring from the mountains and hills. The whole action is changeable as the currents come and go. Material of one texture is often deposited on the top of the material of another texture, thus forming a stratified deposit.

Ammonification: The release of ammonium ions from the decomposing organic materials is known as ammonification. Most microbes can decompose organic matter; ammonification therefore takes place when conditions favour for microbial activity in general.

Anion: Particle with a negative charge.

Anion exchange capacity: Sum of exchangeable anions that a soil can adsorb. Usually expressed as centimoles, or millimoles, of charge per kilogram of soil (or of other adsorbing material such as clay).

Antagonism: Antagonism refers to exceptionally strong competition between pairs of similar ions, such as Ca^{2+} and Sr^{2+}, K^+ and Rb^+, SO_4^{2-} and SeO_4^{2-}, or $H_2PO_4^-$ and $H_2AsO_4^-$.The specific nature of antagonism indicates that the ions involved enter the root the same way whereas other ions may have other entries.

AOAC: It is an abbreviation for Association of Official Analytical Chemist. It is a non-profit making organization, which serves the need of regulatory and research agencies for analytical works. It has its own elaborate system of selection, verification and approval of analytical methods. The approved methods are incorporated in a book entitled *Official Methods of Analysis* of AOAC.

Aqua Regia: A mixture of one part of concentrated nitric acid and three parts of concentrated hydrochloric acid. An extremely oxidizing solvent capable of even dissolving metals like platinum and gold.

Atom: The smallest existing unit of an element. It consists of nucleus, which contains positively charged particles protons and neutrons of no charge.

Atomic number: The number of protons in the nucleus of an atom.

Atomic weight: The average weight or mass of all the isotopes of an element as determined from the proportion in which they are present in a given element.

Available plant nutrient: Plant nutrient in soil, which can be taken up by plants to meet their requirements. It is measured by extracting with certain reagents.

Available water-holding capacity: The amount of water that a soil can store in a form available for plant use is known as available water-holding capacity of the soil. This is often assumed to be the water held with a capillary potential between 1/3 atmosphere (field capacity) and 15 atmospheres (wilting point).

Base: Bases are water solutions containing more OH^- anions than H^+ cations. The greater the concentration of OH^- ions in solution, the more alkaline (basic) it is.

Base saturation: The percent base saturation of soil is a good measure of how much of the cation exchange capacity is being utilized to store plant nutrients. It is defined as the percentage of the cation-exchange capacity occupied by basic cations (usually Ca^{2+}, Mg^{2+}, K^+ and Na^+; it specifically excludes H^+ and Al^{3+} because they produce acid reactions).

B-horizon: The middle part of the soil profile usually contains more clay and has a brighter colour than the topsoil. This layer is called the B horizon, or subsoil. Part of the clay increase is caused by clay from the A horizon being deposited in the B-horizon, and part of it comes from weathering of silt and sand in the B-horizon to form clay. Thus the B-horizon may accumulate a much higher concentration of clay than either A or C-horizon of the same soil.

Biofertilizer: Biological materials used as a source of supplying plant nutrients through various biological reactions in soils/plants.

Biological test: A technique is used for assessing nutrient status of soil using biological materials.

Bulk density: Soil bulk density is the weight of soil solids per unit volume of total soil. The pore space is a part of the volume of soil measured for bulk density, but the water it contains is driven out by oven drying the soil before it is weighed.

Calcareous soils: A calcareous soil is one that contains lime or other carbonates. It is indicted by effervescence (the release of tiny bubbles of CO_2) when an acid such as dilute HCl is applied to the soil.

Calcium Carbonate Equivalent (CCE): It is the acid-neutralizing capacity of a liming material expressed as a weight percentage of pure $CaCO_3$.

Carbon-to-nitrogen ratio: The carbon-to-nitrogen(C: N) ratio of either plant or soil is determined from chemical analyses by dividing the weight of carbon by the weight of nitrogen present in the material. The C: N ratio is the principal factor determining whether immobilization or mineralization occurs as an organic material decomposes.

Carrier: The material that contain any particular plant nutrient. For instance, urea, ammonium sulphate, etc. are nitrogen carriers. Single super phosphate (SSP) and triple super phosphate (TSP) are phosphorus carriers. Sulpahte of potash (SOP) and mureate of potash (MOP) are potassium carriers. Zinc sulphate is a zinc carrier.

Catalysis: The process of enhancing the rate of reaction by means of a small quantity of a substance, which itself does not take part in the chemical reaction but accelerate the reaction.

Catena: A toposequence formed entirely in one kind of parent material is called a soil catena (named from the Latin term chain). Several different catenas having different parent materials and vegetation may occur in the same soil association.

Cation: Particle with positive charge; reactions between anions and cations create electrical forces.

Cation exchange: Interchange between a cation in solution and another cation in the boundary layer between the solution and surface of negatively charged material such as clay or organic matter.

Cation exchange capacity (CEC): The capacity of the clay and humus micelles in soil to hold exchangeable cations is known as the cation exchange capacity of the soil. It has traditionally been expressed in milliequivalents of charge per 100 grams of soil, which are equal to centimoles per kilogram in the SI system of units. The charge sites must be exposed enough for the cations they attract to be exchanged with other cations from the soil water.

Chelate: Chelates are an important group of both natural and synthetic compounds that can immobilize metallic cations (two well known natural chelates are chlorophyll and haemoglobin). The cations can be replaced by other cations because they are very slightly ionized from the organic chelating agent. Chelates are often used as a micronutrient fertilizers.

Chlorosis: The pale green colour of nutrient deficient plants resulting from a shortage of chlorophyll is called chlorosis. Slow growth accompanies chlorosis because chlorophyll is needed for photosynthesis to produce carbohydrates.

Clay: Soil particle smaller than 0.002 mm or 2 µm, with high specific area mainly influencing soil colloidal properties as well as stability of soil structure: high stability in both wet and dry conditions; also a soil texture class.

Clay loam: Soil texture class. See also **soil texture.**

Clay minerals: Clay-sized hydrous aluminium silicates having a large interlayer space that can hold significant amounts of water and other substances; they have a large surface area allowing **swelling** and **shrinking;** examples are montmorillonite or smectite and kaolinite.

Colloid: Particle, which may be a molecular aggregate, with a diameter of 0.1 to 0.001 mm; clay and soil organic matter are often called soil colloids because they have particle sizes that are within, or approach, colloidal dimensions.

Conditioner: A material added to fertilizer to prevent caking and to keep it free flowing.

Critical nutrient level: The concentration level of any nutrient below, which plants show deficiency symptoms and would respond to the application of the nutrient. Critical levels are established regarding concentration of different nutrients in levels of different plants. Critical nutrient level is also marked for nutrient content in soil as extracted by different extractants.

Crust: A soil crust is a common form of compact, excessively strong soil, often resulting from the pounding action of raindrops falling on bare soil. Crusts are denser and stronger than the soil below and contain more humus, clay, and silt than the underlying soil. Many soil crusts are too strong for seedlings to break through; the seedlings are deflected downward and die without emerging from the soil.

Deficiency symptom: Discernable evidence of abnormal growth resulting from a nutrient shortage is a deficiency symptom. The most common plant nutrient deficiency symptoms are stunted growth and chlorosis.

Deverda's Alloy: An alloy containing 50 % copper, 45 % aluminium and 5 % zinc. A strong reducing agent in alkaline media, it is used to convert the nitrate nitrogen into ammonical nitrogen while estimating nitrogen in fertilizers/soils.

Dispersion: The separation of soil aggregates into individual particles is called dispersion. A sodium compound is added to disperse the soil colloids before a mechanical analysis is done.

E horizon: Most forested soils (and some other soils) have an eluviated horizon between the A and B-horizons or above a layer with low permeability. E horizons are lighter coloured than the A above and the B below because eluviations has removed much material from them and left them low in organic matter and other colouring agents.

Electrical conductivity (EC): Conduction of electricity through water or a solution of soil commonly used to estimate the soluble salt content in solution, e.g. soil solution.

Eluviation: The removal of solid material from one horizon (an A or E horizon) is called eluviation, and its deposition in another (a B horizon) is called illuviation. The material may move either as solid particles or in solution.

Epipedon: Diagnostic horizons that necessarily occur at or near the soil surface are called epipedons. Epipedons include but are not limited to A horizons. Underlying E and even B-horizons that fulfill the definitions are included as part of the epipedon.

Equivalent weight: The weight of an element that will combine with 7.9997 of oxygen or 1.0079 of hydrogen. Equivalent weight of molecules is calculated by dividing molecular weight with the number of hydrogen cations or protons/electrons that will take part in any reaction with the molecule, e.g. equivalent weight of sulphuric acid = 98/2 = 49 where 98 is the molecular weight of sulphuric acid and 2 is the number of hydrogen cations that will take part in any reaction with sulphuric acid.

Erodibility: The susceptibility of a soil to the action of erosive forces is called its erodibility. The K factor in the Universal Soil Loss Equation is intended to evaluate soil erodibility.

Erosivity: The erosive capacity of water or wind is evaluated as erosivity. The R factor in the Universal Soil Loss Equation is a measure of the erosivity of rainfall.

Essential elements: The elements without which the plant cannot complete its vegetative or reproductive cycle. There are 17 essential elements for plants.

Eutrophication: The process by which a body of water is enriched with deficient nutrients (usually phosphorus and/or nitrogen) is called eutrophication. The connotation is usually negative because the added nutrients make it possible for unwanted algae to grow in the water.

Fertigation: The practice of adding fertilizer to irrigation water is known as fertigation. Fertigation is most often used with either trickle or sprinkler irrigation because it is harder to get uniform distribution with surface irrigation.

Fertilization: Application of mainly mineral compounds, in order to increase soil fertility. In some cases, (e.g. liming) the purpose of fertilization is also to improve specific soil properties (pH, stability of soil structure).

Fertilizer: Fertilizers are sources of plant nutrients that can be added to soil to supplement its natural fertility. They are intended to supply plant needs directly rather than the indirect manner of soil amendments.

Fertilizer grade: It is the minimum guaranty of plant nutrients in a fertilizer expressed by numerical figures.

Field capacity: Field capacity has been defined as the soil moisture state when, 48 hours after saturation or heavy rain, all downward movement of water has ceased. It is the water content retained at low suctions (5-33 kPa) depending on soil type, and is the upper limit of plant available water.

Filler: Material added to a fertilizer mixture or fertilizer material to make up the difference between the weight of an added ingredient required and the total fertilizer that should be made according to the fertilizer formula. Materials commonly used as fillers are sand, dolomitic limestone, gypsum, furfural, vermiculite, rock salt, coal ashes, chalk grit, spent foundry sand, peanut hulls, saw dust and rice husk, etc.

Fine texture: (i) A broad group of textures consisting of, or containing, large quantities of fine fractions, particularly silt and clay. Includes sandy clay, silty clay, and clay texture classes. (ii) When used in reference to family particle-size classes in U.S. and FAO soil taxonomy, is specifically defined as having 35 to 60 percent clay.

Hallow Cathode Lamp: A light source capable of producing a line spectra of a chemical elements, which is extensively used in atomic absorption spectrophotometers. A typical hallow cathode lamp consists of a cathode contained in a sealed glass envelop and a suitable anode. The sealed envelope contains inert gas like argon or neon at a very low pressure. When high voltage (about 600V) is applied the positively charged gas bombards the cathode and dislodges the atoms, which are exited and a spectrum of chemical elements is emitted. The chemical elements used in a cathode lamp are either a pure element or a mixture of elements. The emission line spectra produce by the lamp containing any particular element, say; zinc in the cathode can be used of an atomic absorption study of that very element. Generally, the life of a cathode lamp is about 1000 hours of operation.

Humus: The relatively stable portion of soil organic matter is called humus. The humus from any soil is a complex mixture of mostly finely divided nonliving organic materials that defy detailed chemical analysis.

Hydromorphic soils: Formed under conditions of poor drainage in marshes, swamps, seepage areas or flats.

Illuviation: The process of deposition of materials moved by eluviations is called illuviation. The horizon where it occurs is called illuvial horizon and normally becomes a B-horizon.

Immobilization: The process by which either plants or microbes convert inorganic ions into organic compounds is called as immobilization.

Indicators: Organic substances which change colour to mark the end of titration with reference to pH, redox potential or precipitation. Depending on the types of reaction and nature of indicators, these are grouped as (i) acid base indicators, (ii) mixed indicators, (iii) universal indicators, (iv) fluorescent indicators, (v) chemiluminescent indicators, (vi) adsorption indicators, and (vii) oxidation-reduction indicators.

Ion: Electrically charged atom or group of atoms.

Lime: Lime is generally defined as calcium oxide (CaO), but in agriculture lime usually means any material used as a soil amendment to raise the soil pH. Ground limestone (mostly $CaCO_3$) is by far the most common agricultural liming material.

Map scale: Relationship between a certain distance on the map and the corresponding distance on the ground (e.g. 1:10,000, which means 1 cm on the map equals to 10,000 cm or 100 m on the ground); the scale is usually located in the legend box of a map.

Mineralization: The process by which organic matter is decomposed and inorganic ions released is called mineralization.

Moderately fine textured: Texture group consisting of clay loam, sandy clay loam and silty clay loam textures.

Munsell Colour System: Colour designation system that specifies the relative degrees of the three simple variables of colour: hue (wavelength), value (degree of lightness or darkness), and chroma (purity or strength). For example: 10YR 6/4 is a colour (of soil) with a hue = 10YR, value = 6, and chroma = 4.

Organic soil: Consists of organic debris that accumulates at the surface under either wet or dry conditions and in which any mineral component present does not significantly affect the soil properties. Organic soil material must have organic carbon (organic matter) contents as follows: (1) if saturated with water for long periods (unless artificially drained), and excluding live roots, either: 18 % organic carbon (30 % organic matter) or more if the mineral fraction comprise 60 % or more clay; or 12 % organic carbon (20 % organic matter) or more if the mineral fraction has no clay; or a proportional lower limit of organic carbon content between 12 and 18 % if the clay content of the mineral fraction is between 0 and 60 %; or (2) if never saturated with water for more than a few days, 20 % or more organic carbon.

Organic soil: A soil in which the sum of the thicknesses of layers comprising organic soil materials is generally greater than the sum of the thicknesses of mineral layers.

Osmotic potential: The osmotic potential results from the materials dissolved in the soil water. Dissolved ions and molecules attract water molecules through hydrogen bonding and thus reduce the free energy of water. The soil solution is very dilute in most soils, so the osmotic potential is commonly very small. But some soils in arid regions contain large concentrations of soluble salts and are called saline soils. The osmotic potential in some saline soils is stronger (more negative) than the gravitational and capillary potentials.

Parent material: Mineral or rock material on and/or from which soils are formed during pedogenesis (soil formation process); parent material is one of the five major soil forming factors.

Particle density: Particle density is the average density of the soil particles. The volume of the soil solids excluding pore space divides the oven-dry weight of the soil. The particle density of a typical soil is often assumed to be 2.65 g/cc.

Peat: Organic soil material with more than 50% of organic matter derived from plant residues with not fully destroyed structure. Peat forms in a wet soil environment or below the water table where mineralization of organic matter comes close to zero; a peat horizon or layer is normally more than 30 cm thick.

Plasticity: Plasticity is the ability to take and hold a new shape when pressure is applied and removed. A nonplastic soil will fall apart when confining pressure is removed. Plastic soils may be classed as slightly plastic, plastic, or very plastic depending on the pressure required to deform the wet soil. Increasing plasticity is correlated with increasing clay content.

Primary mineral: A mineral that has not been altered chemically since crystallization and deposition from molten lava.

Saline soil: A non-*sodic soil* containing sufficient soluble salt to adversely affect the growth of most crop plants. The lower limit of electrical conductivity in the saturation extract of such soils is conventionally set at 4 dS/m (at 25°C), though sensitive plants are affected at about half this salinity and highly tolerant ones at about twice this salinity.

Saline-sodic soil: Salt-affected soils with a high exchangeable sodium percentage (ESP) greater than 15%, pH usually less than 8.5; in general these soils are not suitable for agriculture.

Sand: Soil particles between 0.05 mm and 2 mm with low specific area and also used as a texture class name for coarse soil materials. Unlike clays, sandy soils do not shrink and swell on drying and wetting and, unless artificially compacted, are rapidly permeable.

Salt-affected soil: Soil that has been adversely affected by the presence of soluble salts, with or without high amounts of exchangeable sodium.

Silt: Soil particles between 0.002 mm and 0.05 mm with high or medium-high specific area influencing stability of soil structure; also used as a texture class name for medium and medium-fine soil materials.

Sodic soil: Soil with excess of sodium, pH is higher than 7, usually in the range 8 - 10, exchangeable sodium percentage, ESP >15 and very poor soil structure. These soils need special management and are not used for agriculture; non-*sodic soils* are without excess of sodium.

Soil amendment: A material is called a soil amendment if it is added to alter the chemical or physical properties of a soil. It may also have fertilizer value, but it is added primarily to alter the pH, tilth, or other soil properties. The rates applied are large compared to fertilizer rates, and are usually measured in tons per hectare.

Soil biology: A scientific discipline dealing with living components of soils, which are represented mainly by bacteria, fungi, protozoa, nematodes, arthropods and earthworms as well as by mammals.

Soil chemistry: A scientific discipline dealing with chemical properties of soils and studies on the influence of fertilizers, pesticides and the other chemical substances applied on or into the soil on soil behaviour and fertility.

Soil classification: Also termed soil taxonomy, is the scientific discipline dealing with grouping of soils into soil morphological units or soil types, according to similar or comparable soil forming properties. Many countries in the world have national soil classification systems but those of FAO, WRB and USDA are used internationally. For transnational comparisons, an international soil classification system, into which the majority of national systems can be translated, is needed. In future, this will be the WRB.

Soil colour: Soil colour is one of the indicators of soil status and depends on many factors, mainly on the amount and state of organic matter and iron oxide, as well as amount of air and water in soil pores. In general, dark soils have high organic matter content, grey soils are waterlogged or anaerobic, brown soils are well-drained and aerated soils. Soil colour is measured using *Munsell Soil Color charts.*

Soil compaction: Changing the nature of the soil such that there is a decrease in the volume of voids between soil particles or aggregates; it is manifest as an increase in bulk density and a severely compacted soil can become effectively impermeable. Some soils are naturally compacted, e.g. very heavy textured soils (fine textured). Man-made compaction is caused by the passage of heavy machinery and very intensive soil exploitation.

Soil cracks: Openings in horizontal (mm or several cm) and vertical (cm or several m) orientation, mainly affecting soil hydraulic properties, shrinking processes. Heavy clay soils are more susceptible to cracks formation than loamy soils whereas in sandy soils cracks do not form or they are very small and unstable. Soil cultivation destroys crack system, mainly by tillage.

Soil degradation: Negative process often accelerated by human activities (improper soil use and cultivation practices, building areas) that leads to deterioration of soil properties and functions or destruction of soil as a whole, e.g. compaction, erosion, salinisation, acidification.

Soil depth: Depth of soil profile from the top to parent material or bedrock or to the layer of obstacles for roots. It differs significantly for different soil types. It is one of basic critera used in soil classification. Soils can be very shallow (less than 25 cm), shallow (25 cm-50 cm), moderately deep (50 cm-90 cm), deep (90cm-150 cm) and very deep (more than 150 cm).

Soil fertility: A measure of the ability of soil to provide plants with sufficient amount of nutrients and water, and a suitable medium for root development to assure proper plant growth and maturity.

Soil geography: Scientific discipline dealing with distribution of soil types in landscapes, describing this distribution according to geographical rules.

Soil microorganisms: Represented by protozoa, viruses, bacteria, fungi and algae. The most prevalent are bacteria and fungi, and depending on conditions (water and nutrients content, temperature, etc.) they can be in an active or non-active state. According to nutrient (and oxygen) demand, microorganisms are divided to autotrophic and heterotrophic, (aerobic and anaerobic) groups. Micro-organisms are a good indicator of soil status and quality.

Soil monitoring: Repeated observation and measurement of selected soil properties and functions, mainly for studying changes in soil conditions.

Soil morphology: Form and arrangement of pedological features.

Soil organic matter: The organic fraction of the soil exclusively comprising undecayed plant and animal residues.

Soil physics: Scientific discipline dealing with physical properties of soil (density, porosity, water retention and permeability, hydraulic conductivity, etc.).

Soil porosity: Volume of water and air that can be held in a soil; ratio of the volume of voids to the total volume of the soil.

Soil profile: Vertical section of soil horizons from upper layer to the parent material, showing the arrangement (configuration) of soil horizons typical for single soil types and used as a basis for soil classification.

Soil sorption: Selective process, which occurs on soil particles smaller than 0.002mm (<2 μm); these small particles have colloidal properties, are able to hold and exchange ions, water or gases.

Soil texture: Numerical proportion (% by wt.) of sand, silt and clay in a soil. Sand, silt and clay content are estimated in the field, and/or quantitatively in the laboratory, and then placed within the texture triangle to determine soil texture class. Texture can be coarse (sand particles predominate), medium (silt particles predominate), or fine (clay particles predominate).

Stoniness: It is the relative proportion (vol %) of coarse particles (larger than 2 mm diameter) in the soil or on soil surface; 15% stones is a high value and can hinder cultivation and reduce water holding capacity.

Surface soil: The layer of soil occurring on the surface, synonym *topsoil*.

Swamp: Seasonally flooded low land. Similar to *marsh*, but with more woody plants and to *bog* but with better drainage.

Swelling and shrinking: Two opposite processes of soil volume change. Swelling, increase of soil volume and shrinking, decrease of soil volume. These processes are influenced by actual water content and presence of clay minerals, which are able to take or to lose water in their interlayer spaces. Difference in volume can range from 5% to more than 100% depending on quality and quantity of clay minerals.

Tissue Analysis: Tissue analysis is a laboratory determination of the total content of specified elements (normally plant nutrients) in plant tissue. The plant material selected is dried and finely ground before being tested. Its nutrient contents are then compared with the nutrient levels in the same parts of normal plants at the same stage of maturity. The comparison is used to determine whether there is a nutrient deficiency, or if the supply is adequate or excess for each of the nutrient tested.

Topsoil: (i) The surface soil horizon (A) which is modified when cultivated, and designated as Ap. (ii) Fertile soil material used to top dress road banks, gardens, and lawns.

Water retention: The ability of soil to hold water for a period that is longer than infiltration, normally 48 h in a freely draining soil. It strongly depends on organic matter and bulk density. Soil texture also has an influence on water retention.

Water table: The upper surface of ground water or that level in the ground where the water is at atmospheric pressure.

Weathering: The breakdown and changes in rocks and sediments at or near the earth's surface produced by biological, chemical and physical agents or combinations of them.

Wilting point: Soil moisture content when the rate of absorption of water by plant roots is too slow to maintain plant turgidity and permanent wilting occurs. The average moisture tension at the outside surface of the moisture film around soil particles when permanent wilting occurs is 15 atmospheres or 1500 kPa.

—•—

FSC
www.fsc.org
MIX
Papier | Fördert
gute Waldnutzung
FSC® C083411

Zeitfracht Medien GmbH
Ferdinand-Jühlke-Straße 7
99095 Erfurt, Deutschland
produktsicherheit@kolibri360.de